Brief Introduction to Linear Algebra

STANLEY I. GROSSMAN

Wadsworth Publishing Company
Belmont, California
A Division of Wadsworth, Inc.

Contents

CHAPTER 1 Systems of Linear Equations

    1.1 Introduction, 1
    1.2 Two Linear Equations in Two Unknowns, 2
    1.3 m Equations in n Unknowns: Gauss–Jordan and Gaussian Elimination, 6
    1.4 Homogeneous Systems of Equations, 22
    Review Exercises for Chapter 1, 25

CHAPTER 2 Vectors and Matrices

    2.1 Vectors, 27
    2.2 The Scalar Product of Two Vectors, 33
    2.3 Matrices, 37
    2.4 Matrix Products, 41
    2.5 Matrices and Linear Systems of Equations, 50
    2.6 Linear Independence and Homogeneous Systems, 53
    2.7 The Inverse of a Square Matrix, 59
    2.8 The Transpose of a Matrix, 77
    Review Exercises for Chapter 2, 80

CHAPTER 3 Determinants

    3.1 Definitions, 82
    3.2 Properties of Determinants, 88
    3.3 If Time Permits: Proof of the Basic Theorem, 103
    3.4 Determinants and Inverses, 105
    3.5 Cramer's Rule, 112
    Review Exercises for Chapter 3, 116

    Answers to Odd-Numbered Problems   118

    Index   126

© 1984, 1980 by Wadsworth, Inc. All rights reserved. No part of this book may be reproduced, stored in a retrieval system, or transcribed, in any form or by any means, electronic, mechanical, photocopying, recording, or otherwise, without the prior written permission of the publisher, Wadsworth Publishing Company, Belmont, California 94002, a division of Wadsworth, Inc.

Printed in the United States of America

1 2 3 4 5 6 7 8 9 10—88 87 86 85 84   ISBN 0-534-03495-0

1 Systems of Linear Equations

1.1 Introduction

This is a book about linear algebra. If you look up the word "linear" in a dictionary, you will find something like the following: lin-e-ar (lin' ē ər), adj. 1. of, consisting of, or using lines.† In mathematics, the word "linear" means a good deal more than that. Nevertheless, much of the theory of elementary linear algebra is in fact a generalization of properties of straight lines. As a review, here are some fundamental facts about straight lines:

i. The **slope** m of a line passing through the points (x_1, y_1) and (x_2, y_2) is given by (if $x_2 \neq x_1$)

$$m = \frac{y_2 - y_1}{x_2 - x_1} = \frac{\Delta y}{\Delta x}$$

ii. If $x_2 - x_1 = 0$ and $y_2 \neq y_1$, then the line is vertical and the slope is said to be **undefined**.‡
iii. Any line (except one with infinite slope) can be described by writing its equation in the slope-intercept form $y = mx + b$, where m is the slope of the line and b is the y-intercept of the line (the value of y at the point where the line crosses the y-axis).
iv. Two lines are parallel if and only if they have the same slope.
v. If the equation of a line is written in the form $ax + by = c$ ($b \neq 0$), then, as is easily computed, $m = -a/b$.
vi. If m_1 is the slope of line L_1, m_2 is the slope of line L_2, $m_1 \neq 0$, and L_1 and L_2 are perpendicular, then $m_2 = -1/m_1$.
vii. Lines parallel to the x-axis have a slope of zero.
viii. Lines parallel to the y-axis have an undefined slope.

In the next section we shall illustrate the relationship between solving systems of equations and finding points of intersection of pairs of straight lines.

---

† Taken from the pocket edition of *The Random House Dictionary*.
‡ In some textbooks a vertical line is said to have "an infinite slope."

1.2 Two Linear Equations in Two Unknowns

Consider the following system of two linear equations in the two unknowns x_1 and x_2:

$$a_{11}x_1 + a_{12}x_2 = b_1$$
$$a_{21}x_1 + a_{22}x_2 = b_2 \tag{1}$$

where a_{11}, a_{12}, a_{21}, a_{22}, b_1, and b_2 are given numbers. Each of these equations is the equation of a straight line (in the $x_1 x_2$-plane instead of the xy-plane). The slope of the first line is $-a_{11}/a_{12}$; the slope of the second line is $-a_{21}/a_{22}$ (if $a_{12} \neq 0$ and $a_{22} \neq 0$). A **solution** to system (1) is a pair of numbers, denoted (x_1, x_2), that satisfies (1). The questions that naturally arise are whether (1) has any solutions and if so, how many? We shall answer these questions after looking at some examples. In these examples we will make use of two important facts from elementary algebra:

Fact A If $a = b$ and $c = d$, then $a + c = b + d$.

Fact B If $a = b$ and c is any real number, then $ca = cb$.

Fact A states that if we add two equations together, we obtain a third, valid equation. Fact B states that if we multiply both sides of an equation by a constant, we obtain a second, valid equation.

EXAMPLE 1

Consider the system

$$x_1 - x_2 = 7$$
$$x_1 + x_2 = 5 \tag{2}$$

Adding the two equations together gives us, by Fact A, the following equation: $2x_1 = 12$ (or $x_1 = 6$). Then, from the second equation, $x_2 = 5 - x_1 = 5 - 6 = -1$. Thus the pair $(6, -1)$ satisfies system (2) and the way we found the solution shows that it is the only pair of numbers to do so. That is, system (2) has a **unique solution**.

EXAMPLE 2

Consider the system

$$x_1 - x_2 = 7$$
$$2x_1 - 2x_2 = 14 \tag{3}$$

It is apparent that these two equations are equivalent. To see this multiply the first by 2. (This is permitted by Fact B.) Then $x_1 - x_2 = 7$ or $x_2 = x_1 - 7$. Thus the pair $(x_1, x_1 - 7)$ is a solution to system (3) for any real number x_1. That is, system (3) has an **infinite number of solutions**. For example, the following pairs are solutions: $(7, 0)$, $(0, -7)$, $(8, 1)$, $(1, -6)$, $(3, -4)$, and $(-2, -9)$.

EXAMPLE 3 Consider the system

$$x_1 - x_2 = 7$$
$$2x_1 - 2x_2 = 13 \tag{4}$$

Multiplying the first equation by 2 (which, again, is permitted by Fact B) gives us $2x_1 - 2x_2 = 14$. This contradicts the second equation. Thus system (4) has **no solution**.

It is easy to explain, geometrically, what is going on in the preceding examples. First we repeat that the equations in system (1) are both equations of straight lines. A solution to (1) is a point (x_1, x_2) that lies on both lines. If the two lines are not parallel, then they intersect at a single point. If they are parallel, then either they never intersect (no points in common) or they are the same line (infinite number of points in common). In Example 1 the lines have slopes of 1 and -1, respectively. Thus they are not parallel. They have the single point $(6, -1)$ in common. In Example 2 the lines are parallel (slope of 1) and coincident. In Example 3 the lines are parallel and distinct. These relationships are all illustrated in Figure 1.1.

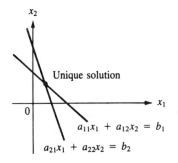
(a) Lines not parallel; one point of intersection

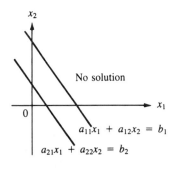
(b) Lines parallel; no points of intersection

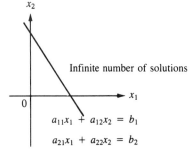
(c) Lines coincide; infinite number of points of intersection

Figure 1.1

Let us now solve system (1) formally. We have

$$a_{11}x_1 + a_{12}x_2 = b_1$$
$$a_{21}x_1 + a_{22}x_2 = b_2$$

Multiplying the first equation by a_{22} and the second by a_{12} yields

$$a_{11}a_{22}x_1 + a_{12}a_{22}x_2 = a_{22}b_1$$
$$a_{12}a_{21}x_1 + a_{12}a_{22}x_2 = a_{12}b_2 \tag{5}$$

Before continuing we note that system (1) and system (5) are **equivalent**. By that we mean that any solution to system (1) is a solution to system (5) and vice versa. This follows immediately from Fact B. Next we subtract the second

equation from the first to obtain
$$(a_{11}a_{22} - a_{12}a_{21})x_1 = a_{22}b_1 - a_{12}b_2 \tag{6}$$

At this point we must pause. If $a_{11}a_{22} - a_{12}a_{21} \neq 0$, then we can divide by it to obtain
$$x_1 = \frac{a_{22}b_1 - a_{12}b_2}{a_{11}a_{22} - a_{12}a_{21}}$$

Then we can "plug" this value of x_1 into system (1) to solve for x_2 and we have found the unique solution to the system. We define the **determinant** of system (1) by

$$\boxed{\text{Determinant of system (1)} = a_{11}a_{22} - a_{12}a_{21}} \tag{7}$$

and we have shown the following:

$$\boxed{\begin{array}{c}\text{If the determinant of system (1)} \neq 0, \text{ then}\\ \text{the system has a unique solution.}\end{array}} \tag{8}$$

How does this statement relate to what we discussed earlier? In system (1) we see that the slope of the first line is $-a_{11}/a_{12}$ and the slope of the second is $-a_{21}/a_{22}$.† In Problems 31, 32, and 33 you are asked to show that the determinant of system (1) is zero if and only if the lines are parallel (have the same slope). So, if the determinant is *not* zero, the lines are not parallel and the system has a unique solution.

We now put the facts discussed above together in a theorem. It is a theorem that will be generalized in later sections of this chapter and in subsequent chapters. We shall keep track of our progress by referring to the theorem as our "Summing Up Theorem." When all its parts have been proved, we shall see a remarkable relationship among several important concepts in linear algebra.

THEOREM 1 **SUMMING UP THEOREM—VIEW 1** The system
$$a_{11}x_1 + a_{12}x_2 = b_1$$
$$a_{21}x_1 + a_{22}x_2 = b_2$$

of two equations in the two unknowns x_1 and x_2 has no solution, a unique solution, or an infinite number of solutions. It has:

i. A unique solution if and only if its determinant is not zero.
ii. No solution or an infinite number of solutions if and only if its determinant is zero.

---

† Here we are assuming that neither a_{12} nor a_{22} equals zero. If either a_{12} or a_{22} is zero, then system (1) can be easily solved. If $a_{12} = 0$, for example, then $x_1 = b_1/a_{11}$. Thus we shall not worry about this possibility.

In the next section we will discuss systems of m equations in n unknowns and will see that always there is either no solution, one solution, or an infinite number of solutions. In Chapter 3 we define and calculate determinants for systems of n equations in n unknowns and shall find that our Summing Up Theorem—Theorem 1—is true in this general setting.

PROBLEMS 1.2

In Problems 1–12, find all solutions (if any) to the given systems. In each case calculate the determinant.

1. $x_1 - 3x_2 = 4$
   $-4x_1 + 2x_2 = 6$
2. $2x_1 - x_2 = -3$
   $5x_1 + 7x_2 = 4$
3. $2x_1 - 8x_2 = 5$
   $-3x_1 + 12x_2 = 8$
4. $2x_1 - 8x_2 = 6$
   $-3x_1 + 12x_2 = -9$
5. $6x_1 + x_2 = 3$
   $-4x_1 - x_2 = 8$
6. $3x_1 + x_2 = 0$
   $2x_1 - 3x_2 = 0$
7. $4x_1 - 6x_2 = 0$
   $-2x_1 + 3x_2 = 0$
8. $5x_1 + 2x_2 = 3$
   $2x_1 + 5x_2 = 3$
9. $2x_1 + 3x_2 = 4$
   $3x_1 + 4x_2 = 5$
10. $ax_1 + bx_2 = c$
    $ax_1 - bx_2 = c$
11. $ax_1 + bx_2 = c$
    $bx_1 + ax_2 = c$
12. $ax_1 - bx_2 = c$
    $bx_1 + ax_2 = d$

13. Find conditions on a and b such that the system in Problem 10 has a unique solution.
14. Find conditions on a, b, and c such that the system in Problem 11 has an infinite number of solutions.
15. Find conditions on a, b, c, and d such that the system in Problem 12 has no solutions.

In Problems 16–21, find the point of intersection (if there is one) of the two lines.

16. $x - y = 7$; $2x + 3y = 1$
17. $y - 2x = 4$; $4x - 2y = 6$
18. $4x - 6y = 7$; $6x - 9y = 12$
19. $4x - 6y = 10$; $6x - 9y = 15$
20. $3x + y = 4$; $y - 5x = 2$
21. $3x + 4y = 5$; $6x - 7y = 8$

Let L be a line and let $L_\perp$ denote the line perpendicular to L that passes through a given point P. The **distance** from L to P is defined to be the distance† between P and the point of intersection of L and $L_\perp$. In Problems 22–27, find the distance between the given line and point.

22. $x - y = 6$; $(0, 0)$
23. $2x + 3y = -1$; $(0, 0)$
24. $3x + y = 7$; $(1, 2)$
25. $5x - 6y = 3$; $(2, \frac{16}{5})$
26. $2y - 5x = -2$; $(5, -3)$
27. $6y + 3x = 3$; $(8, -1)$
28. Find the distance between the line $2x - y = 6$ and the point of intersection of the lines $2x - 3y = 1$ and $3x + 6y = 12$.
*29. Prove that the distance between the point (x_1, y_1) and the line $ax + by = c$ is given by
$$d = \frac{|ax_1 + by_1 - c|}{\sqrt{a^2 + b^2}}$$

30. A zoo keeps birds (two-legged) and beasts (four-legged). If the zoo contains 60 heads and 200 feet, how many birds and how many beasts live there?
31. Suppose that the determinant of system (1) is zero. Show that the lines given in (1) are parallel.

---

† Recall that if (x_1, y_1) and (x_2, y_2) are two points in the xy-plane, then the distance d between them is given by $d = \sqrt{(x_1 - x_2)^2 + (y_1 - y_2)^2}$.

32. If there is a unique solution to system (1), show that its determinant is nonzero.

33. If the determinant of system (1) is nonzero, show that the system has a unique solution.

34. The Sunrise Porcelain Company manufactures ceramic cups and saucers. For each cup or saucer a worker measures a fixed amount of material and puts it into a forming machine, from which it is automatically glazed and dried. On the average, a worker needs 3 minutes to get the process started for a cup and 2 minutes for a saucer. The material for a cup costs 25¢ and the material for a saucer costs 20¢. If $44 is allocated daily for production of cups and saucers, how many of each can be manufactured in an 8-hour work day if a worker is working every minute and exactly $44 is spent on materials?

35. Answer the question of Problem 34 if the materials for a cup and saucer cost 15¢ and 10¢, respectively, and $24 is spent in an 8-hour day.

36. Answer the question of Problem 35 if $25 is spent in an 8-hour day.

37. An ice-cream shop sells only ice-cream sodas and milk shakes. It puts 1 ounce of syrup and 4 ounces of ice cream in an ice-cream soda, and 1 ounce of syrup and 3 ounces of ice cream in a milk shake. If the store used 4 gallons of ice cream and 5 quarts of syrup in a day, how many ice-cream sodas and milk shakes did it sell? [Hint: 1 quart = 32 ounces; 1 gallon = 128 ounces]

1.3 m Equations in n Unknowns: Gauss-Jordan and Gaussian Elimination

In this section we describe a method for finding all solutions (if any) to a system of m linear equations in n unknowns. In doing so we shall see that, like the 2×2 case, such a system has no solutions, one solution, or an infinite number of solutions. Before launching into the general method, let us look at some simple examples.

EXAMPLE 1 Solve the system

$$2x_1 + 4x_2 + 6x_3 = 18$$
$$4x_1 + 5x_2 + 6x_3 = 24 \qquad (1)$$
$$3x_1 + x_2 - 2x_3 = 4$$

Solution Here we seek three numbers x_1, x_2, and x_3 such that the three equations in (1) are satisfied. Our method of solution will be to simplify the equations as we did in the last section so that solutions can be readily identified. We begin by dividing the first equation by 2. This gives us

$$x_1 + 2x_2 + 3x_3 = 9$$
$$4x_1 + 5x_2 + 6x_3 = 24 \qquad (2)$$
$$3x_1 + x_2 - 2x_3 = 4$$

As we saw in the last section (Fact A), adding two equations together leads to a third, valid equation. This equation may replace either of the two equations used to obtain it in the system. We begin simplifying system (2) by multiplying both sides of the first equation in (2) by -4 and adding this new equation to the second equation. This gives us

$$-4x_1 - 8x_2 - 12x_3 = -36$$
$$4x_1 + 5x_2 + 6x_3 = 24$$
$$\overline{ - 3x_2 - 6x_3 = -12}$$

The equation $-3x_2 - 6x_3 = -12$ is our new second equation and the system is now

$$x_1 + 2x_2 + 3x_3 = 9$$
$$-3x_2 - 6x_3 = -12$$
$$3x_1 + x_2 - 2x_3 = 4$$

We then multiply the first equation by -3 and add it to the third equation:

$$x_1 + 2x_2 + 3x_3 = 9$$
$$-3x_2 - 6x_3 = -12 \qquad (3)$$
$$-5x_2 - 11x_3 = -23$$

Note that in system (3) the variable x_1 has been eliminated from the second and third equations. Next we divide the second equation by -3:

$$x_1 + 2x_2 + 3x_3 = 9$$
$$x_2 + 2x_3 = 4$$
$$-5x_2 - 11x_3 = -23$$

We multiply the second equation by -2 and add it to the first and then multiply the second equation by 5 and add it to the third:

$$x_1 - x_3 = 1$$
$$x_2 + 2x_3 = 4$$
$$- x_3 = -3$$

We multiply the third equation by -1:

$$x_1 - x_3 = 1$$
$$x_2 + 2x_3 = 4$$
$$x_3 = 3$$

Finally, we add the third equation to the first and then multiply the third equation by -2 and add it to the second to obtain the following system (which is equivalent to system (1)):

$$\begin{aligned} x_1 &= 4 \\ x_2 &= -2 \\ x_3 &= 3 \end{aligned}$$

This is the unique solution to the system. We write it in the form $(4, -2, 3)$. The method we used here is called **Gauss-Jordan elimination**.†

Before going on to another example, let us summarize what we have done in this example:

i. We divided to make the coefficient of x_1 in the first equation equal to 1.
ii. We "eliminated" the x_1 terms in the second and third equations. That is, we made the coefficients of these terms equal to zero by multiplying the first equation by appropriate numbers and then adding it to the second and third equations, respectively.
iii. We divided to make the coefficient of the x_2 term in the second equation equal to 1 and then proceeded to use the second equation to eliminate the x_2 terms in the first and third equations.
iv. We divided to make the coefficient of the x_3 term in the third equation equal to 1 and then proceeded to use the third equation to eliminate the x_3 terms in the first and second equations.

We emphasize that, at every step, we obtained systems that were equivalent. That is, each system had the same set of solutions as the one that preceded it. This follows from Facts A and B on page 2.

Before solving other systems of equations, we introduce notation that makes it easier to write down each step in our procedure. A **matrix** is a rectangular array of numbers. For example, the coefficients of the variables x_1, x_2, x_3 in system (1) can be written as the entries of a matrix A, called the **coefficient matrix** of the system:

$$A = \begin{pmatrix} 2 & 4 & 6 \\ 4 & 5 & 6 \\ 3 & 1 & -2 \end{pmatrix} \qquad (4)$$

The study of matrices will take a large part of the remaining chapters of this text. We introduce them here for convenience of notation.

Using matrix notation, system (1) can be written as the **augmented matrix**

$$\begin{pmatrix} 2 & 4 & 6 & | & 18 \\ 4 & 5 & 6 & | & 24 \\ 3 & 1 & -2 & | & 4 \end{pmatrix} \qquad (5)$$

---

† Named after the great German mathematician Karl Friedrich Gauss (1777–1855) and the French mathematician Camille Jordan (1838–1922).

Carl Friedrich Gauss
(Library of Congress)

A man of awesome mathematical stature and talent, Carl Friedrich Gauss straddled the eighteenth and nineteenth centuries like a mathematical Colossus of Rhodes. He is universally regarded as the greatest mathematician of the nineteenth century and, along with Archimedes and Isaac Newton, as one of the three greatest mathematicians of all time.

Carl was born in Brunswick, Germany, in 1777. His father was a hardworking laborer with stubborn and unappreciative views of formal education. His mother, however, though uneducated herself, encouraged the boy in his studies and maintained a lifelong pride in her son's achievements.

Carl was one of those remarkable infant prodigies who appear from time to time. They tell of him the incredible story that at the age of three he detected an arithmetical error in his father's bookkeeping. And there is the often-told story that when ten years old and in the public schools, his teacher, to keep the class occupied, set the pupils to adding the numbers from 1 to 100. Almost immediately Carl placed his slate, writing side down, on the annoyed teacher's desk. When all the slates were finally turned in, the amazed teacher found that Carl alone had the correct answer, 5050, but with no accompanying calculation. Carl had mentally summed the arithmetic progression $1 + 2 + 3 + \cdots + 98 + 99 + 100$ by noting that $100 + 1 = 101$, $99 + 2 = 101$, $98 + 3 = 101$, and so on for 50 such pairs, whence the answer is 50×101, or 5050. Later in life Gauss used to claim jocularly that he could figure before he could talk.

Gauss's precocity came to the attention of the Duke of Brunswick, who, as a kindly and understanding patron, saw the boy enter the college in Brunswick at the age of 15, and then Göttingen University at the age of 18. Vacillating between becoming a philologist or a mathematician (though he had already devised the method of least squares a decade before Legendre independently published it), his mind was dramatically made up in favor of mathematics on March 30, 1796, when he was still a month short of his nineteenth birthday. The event was his surprising contribution to the theory of the Euclidean construction of regular polygons and, in particular, the discovery that a regular polygon of 17 sides can be so constructed.

In his doctorial dissertation, at the University of Helmstädt and written at the age of twenty, Gauss gave the first wholly satisfactory proof of the *fundamental theorem of algebra* (that a polynomial equation with complex coefficients and of degree n has at least one complex root). Unsuccessful attempts to prove this theorem had been made by Newton, Euler, d'Alembert, and Lagrange.

Gauss contributed noteworthily to astronomy, geodesy, and electricity. In 1801 he calculated, by a new procedure and from meager data, the orbit of the then recently discovered planetoid Ceres, and in the following year that of the planetoid Pallas. In 1807 he became professor of mathematics and director of the observatory at Göttingen, a post that he held until his death. In 1821 he carried out a triangulation of Hannover, measured a meridional arc, and invented the heliotrope (or heliograph). In 1831 he commenced collaboration with his colleague Wilhelm Weber (1804–1891) in basic research in electricity and magnetism, and in 1833 the two scientists devised the electromagnetic telegraph.

Famous is Gauss's assertion that "mathematics is the queen of the sciences, and the theory of numbers is the queen of mathematics." Gauss has been described as "the mathematical giant who from his lofty heights embraces in one view the stars and the abysses." In his scientific writing, Gauss was a perfectionist. Claiming that a cathedral is not a cathedral until the last piece of scaffolding is removed, he strove to make each of his works complete, concise, polished, and convincing, with every trace of the analysis by which he reached his results removed. He accordingly adopted as his seal a tree bearing only a few fruits and carrying the motto: *Pauca sed matura* (*Few, but ripe*). Gauss chose for his second motto the following lines from *King Lear*:

Thou, nature, art my goddess; to thy laws
My services are bound . . . .

Gauss thus believed that mathematics, for inspiration, must touch the real world. As Wordsworth put it, "Wisdom oft is nearer when we stoop than when we soar."

Gauss died in his home at the Göttingen Observatory in 1855.

For example, the first row in the augmented matrix (5) is read $2x_1 + 4x_2 + 6x_3 = 18$. Note that each row of the augmented matrix corresponds to one of the equations in the system.

We now introduce some terminology. We have seen that multiplying (or dividing) the sides of an equation by a nonzero number gives us a new, valid equation. Moreover, adding a multiple of one equation to another equation in a system gives us another valid equation. Finally, if we interchange two equations in a system of equations, we obtain an equivalent system. These three operations, when applied to the rows of the augmented matrix representation of a system of equations, are called **elementary row operations**.

To sum up, the three elementary row operations applied to the augmented matrix representation of a system of equations are:

ELEMENTARY ROW OPERATIONS
  i. Multiply (or divide) one row by a nonzero number.
  ii. Add a multiple of one row to another row.
  iii. Interchange two rows.

The process of applying elementary row operations to simplify an augmented matrix is called **row reduction**.

Notation
  i. $M_i(c)$ stands for "multiply the ith row of a matrix by the number c."
  ii. $A_{i,j}(c)$ stands for "multiply the ith row by c and add it to the jth row."
  iii. $P_{i,j}$ stands for "interchange (permute) rows i and j."
  iv. $A \to B$ indicates that the augmented matrices A and B are equivalent; that is, the systems they represent have the same solution.

In Example 1 we saw that by using the elementary row operations (i) and (ii) several times we could obtain a system in which the solutions to the system were given explicitly. We now repeat the steps in Example 1, using the notation just introduced:

$$\begin{pmatrix} 2 & 4 & 6 & | & 18 \\ 4 & 5 & 6 & | & 24 \\ 3 & 1 & -2 & | & 4 \end{pmatrix} \xrightarrow{M_1(\frac{1}{2})} \begin{pmatrix} 1 & 2 & 3 & | & 9 \\ 4 & 5 & 6 & | & 24 \\ 3 & 1 & -2 & | & 4 \end{pmatrix} \xrightarrow{\substack{A_{1,2}(-4) \\ A_{1,3}(-3)}} \begin{pmatrix} 1 & 2 & 3 & | & 9 \\ 0 & -3 & -6 & | & -12 \\ 0 & -5 & -11 & | & -23 \end{pmatrix}$$

$$\xrightarrow{M_2(-\frac{1}{3})} \begin{pmatrix} 1 & 2 & 3 & | & 9 \\ 0 & 1 & 2 & | & 4 \\ 0 & -5 & -11 & | & -23 \end{pmatrix} \xrightarrow{\substack{A_{2,1}(-2) \\ A_{2,3}(5)}} \begin{pmatrix} 1 & 0 & -1 & | & 1 \\ 0 & 1 & 2 & | & 4 \\ 0 & 0 & -1 & | & -3 \end{pmatrix}$$

$$\xrightarrow{M_3(-1)} \begin{pmatrix} 1 & 0 & -1 & | & 1 \\ 0 & 1 & 2 & | & 4 \\ 0 & 0 & 1 & | & 3 \end{pmatrix} \xrightarrow{\substack{A_{3,1}(1) \\ A_{3,2}(-2)}} \begin{pmatrix} 1 & 0 & 0 & | & 4 \\ 0 & 1 & 0 & | & -2 \\ 0 & 0 & 1 & | & 3 \end{pmatrix}$$

Again we can easily "see" the solution $x_1 = 4$, $x_2 = -2$, $x_3 = 3$.

EXAMPLE 2 Solve the system
$$2x_1 + 4x_2 + 6x_3 = 18$$
$$4x_1 + 5x_2 + 6x_3 = 24$$
$$2x_1 + 7x_2 + 12x_3 = 30$$

Solution We proceed as in Example 1, first writing the system as an augmented matrix:
$$\begin{pmatrix} 2 & 4 & 6 & | & 18 \\ 4 & 5 & 6 & | & 24 \\ 2 & 7 & 12 & | & 30 \end{pmatrix}$$

We then obtain, successively,

$$\xrightarrow{M_1(\frac{1}{2})} \begin{pmatrix} 1 & 2 & 3 & | & 9 \\ 4 & 5 & 6 & | & 24 \\ 2 & 7 & 12 & | & 30 \end{pmatrix} \xrightarrow[A_{1,3}(-2)]{A_{1,2}(-4)} \begin{pmatrix} 1 & 2 & 3 & | & 9 \\ 0 & -3 & -6 & | & -12 \\ 0 & 3 & 6 & | & 12 \end{pmatrix}$$

$$\xrightarrow{M_2(-\frac{1}{3})} \begin{pmatrix} 1 & 2 & 3 & | & 9 \\ 0 & 1 & 2 & | & 4 \\ 0 & 3 & 6 & | & 12 \end{pmatrix} \xrightarrow[A_{2,3}(-3)]{A_{2,1}(-2)} \begin{pmatrix} 1 & 0 & -1 & | & 1 \\ 0 & 1 & 2 & | & 4 \\ 0 & 0 & 0 & | & 0 \end{pmatrix}$$

This is equivalent to the system of equations
$$x_1 \quad\quad - x_3 = 1$$
$$x_2 + 2x_3 = 4$$

This is as far as we can go. There are now only two equations in the three unknowns x_1, x_2, x_3 and there are an infinite number of solutions. To see this, let x_3 be chosen. Then $x_2 = 4 - 2x_3$ and $x_1 = 1 + x_3$. This will be a solution for any number x_3. We write these solutions in the form $(1 + x_3, 4 - 2x_3, x_3)$. For example, if $x_3 = 0$ we obtain the solution $(1, 4, 0)$. For $x_3 = 10$ we obtain the solution $(11, -16, 10)$.

EXAMPLE 3 Solve the system
$$2x_1 + 4x_2 + 6x_3 = 18$$
$$4x_1 + 5x_2 + 6x_3 = 24 \qquad (6)$$
$$2x_1 + 7x_2 + 12x_3 = 40$$

Solution We use the augmented-matrix form and proceed exactly as in Example 2 to obtain, successively, the following systems. (Note how, in each step, we use either elementary row operation (*i*) or (*ii*).)

$$\begin{pmatrix} 2 & 4 & 6 & | & 18 \\ 4 & 5 & 6 & | & 24 \\ 2 & 7 & 12 & | & 40 \end{pmatrix} \xrightarrow{M_1(\frac{1}{2})} \begin{pmatrix} 1 & 2 & 3 & | & 9 \\ 4 & 5 & 6 & | & 24 \\ 2 & 7 & 12 & | & 40 \end{pmatrix}$$

$$\xrightarrow[A_{1,3}(-2)]{A_{1,2}(-4)} \begin{pmatrix} 1 & 2 & 3 & | & 9 \\ 0 & -3 & -6 & | & -12 \\ 0 & 3 & 6 & | & 22 \end{pmatrix} \xrightarrow{M_2(-\frac{1}{3})} \begin{pmatrix} 1 & 2 & 3 & | & 9 \\ 0 & 1 & 2 & | & 4 \\ 0 & 3 & 6 & | & 22 \end{pmatrix}$$

$$\xrightarrow[A_{2,3}(-3)]{A_{2,1}(-2)} \begin{pmatrix} 1 & 0 & -1 & | & 1 \\ 0 & 1 & 2 & | & 4 \\ 0 & 0 & 0 & | & 10 \end{pmatrix} \xrightarrow{M_3(\frac{1}{10})} \begin{pmatrix} 1 & 0 & -1 & | & 1 \\ 0 & 1 & 2 & | & 4 \\ 0 & 0 & 0 & | & 1 \end{pmatrix}$$

The last equation now reads $0x_1 + 0x_2 + 0x_3 = 1$, which is impossible since $0 \neq 1$. Thus system (6) has *no* solution.

Let us take another look at these three examples. In Example 1 we began with the matrix

$$A_1 = \begin{pmatrix} 2 & 4 & 6 \\ 4 & 5 & 6 \\ 3 & 1 & -2 \end{pmatrix}.$$

In the process of row reduction A_1 was "reduced" to the matrix

$$R_1 = \begin{pmatrix} 1 & 0 & 0 \\ 0 & 1 & 0 \\ 0 & 0 & 1 \end{pmatrix}.$$

In Example 2 we started with

$$A_2 = \begin{pmatrix} 2 & 4 & 6 \\ 4 & 5 & 6 \\ 2 & 7 & 12 \end{pmatrix}$$

and ended up with

$$R_2 = \begin{pmatrix} 1 & 0 & -1 \\ 0 & 1 & 2 \\ 0 & 0 & 0 \end{pmatrix}.$$

In Example 3 we began with

$$A_3 = \begin{pmatrix} 2 & 4 & 6 \\ 4 & 5 & 6 \\ 2 & 7 & 12 \end{pmatrix}$$

and again ended up with

$$R_3 = \begin{pmatrix} 1 & 0 & -1 \\ 0 & 1 & 2 \\ 0 & 0 & 0 \end{pmatrix}.$$

The matrices R_1, R_2, and R_3 are called the *reduced row echelon forms* of the matrices A_1, A_2, and A_3, respectively. In general, we have the following definition.

DEFINITION 1   **REDUCED ROW ECHELON FORM**   A matrix is in **reduced row echelon form** if the following four conditions hold:

i. All rows (if any) consisting entirely of zeros appear at the bottom of the matrix.
ii. The first nonzero number (starting from the left) in any row not consisting entirely of zeros is 1.
iii. If two successive rows do not consist entirely of zeros, then the first 1 in the lower row occurs farther to the right than the first 1 in the higher row.
iv. Any column containing the first 1 in a row has zeros everywhere else.

EXAMPLE 4   The following matrices are in reduced row echelon form:

i. $\begin{pmatrix} 1 & 0 & 0 \\ 0 & 1 & 0 \\ 0 & 0 & 1 \end{pmatrix}$ ii. $\begin{pmatrix} 1 & 0 & 0 & 0 \\ 0 & 1 & 0 & 0 \\ 0 & 0 & 0 & 1 \end{pmatrix}$ iii. $\begin{pmatrix} 1 & 0 & 0 & 5 \\ 0 & 0 & 1 & 2 \end{pmatrix}$

iv. $\begin{pmatrix} 1 & 0 \\ 0 & 1 \end{pmatrix}$ v. $\begin{pmatrix} 1 & 0 & 2 & 5 \\ 0 & 1 & 3 & 6 \\ 0 & 0 & 0 & 0 \end{pmatrix}$

DEFINITION 2   **ROW ECHELON FORM**   A matrix is in **row echelon form** if conditions (*i*), (*ii*), and (*iii*) hold in Definition 1.

EXAMPLE 5   The following matrices are in row echelon form:

i. $\begin{pmatrix} 1 & 2 & 3 \\ 0 & 1 & 5 \\ 0 & 0 & 1 \end{pmatrix}$ ii. $\begin{pmatrix} 1 & -1 & 6 & 4 \\ 0 & 1 & 2 & -8 \\ 0 & 0 & 0 & 1 \end{pmatrix}$

iii. $\begin{pmatrix} 1 & 0 & 2 & 5 \\ 0 & 0 & 1 & 2 \end{pmatrix}$ iv. $\begin{pmatrix} 1 & 2 \\ 0 & 1 \end{pmatrix}$ v. $\begin{pmatrix} 1 & 3 & 2 & 5 \\ 0 & 1 & 3 & 6 \\ 0 & 0 & 0 & 0 \end{pmatrix}$

Remark 1. The difference between these two forms should be clear from the examples. In row echelon form, all the numbers below the first 1 in a row are zero. In reduced row echelon form, all the numbers above and below the first 1 in a row are zero. Thus reduced row echelon form is more exclusive. That is, every matrix in reduced row echelon form is in row echelon form, but not conversely.

Remark 2. We can always reduce a matrix to reduced row echelon form or row echelon form by performing elementary row operations. We saw this reduction to reduced row echelon form in Examples 1, 2, and 3.

As we saw in Examples 1, 2, and 3, there is a strong connection between the reduced row echelon form of a matrix and the existence of a unique solution to the system. In Example 1, the reduced row echelon form of the **coefficient matrix** (that is, the first three columns of the augmented matrix) had a 1 in each row and there was a unique solution. In Examples 2 and 3, the reduced row echelon form of the coefficient matrix had a row of zeros and the system had either no solution or an infinite number of solutions. This turns out always to be true in any system with the same number of equations as unknowns. But before turning to the general case, let us discuss the usefulness of the row echelon form of a matrix. It is possible to solve the system in Example 1 by reducing the coefficient matrix to its row echelon form.

EXAMPLE 6 Solve the system of Example 1 by reducing the coefficient matrix to row echelon form.

Solution We begin as before:

$$\begin{pmatrix} 2 & 4 & 6 & | & 18 \\ 4 & 5 & 6 & | & 24 \\ 3 & 1 & -2 & | & 4 \end{pmatrix} \xrightarrow{M_1(\frac{1}{2})} \begin{pmatrix} 1 & 2 & 3 & | & 9 \\ 4 & 5 & 6 & | & 24 \\ 3 & 1 & -2 & | & 4 \end{pmatrix} \xrightarrow[A_{1,3}(-3)]{A_{1,2}(-4)}$$

$$\begin{pmatrix} 1 & 2 & 3 & | & 9 \\ 0 & -3 & -6 & | & -12 \\ 0 & -5 & -11 & | & -23 \end{pmatrix} \xrightarrow{M_2(-\frac{1}{3})} \begin{pmatrix} 1 & 2 & 3 & | & 9 \\ 0 & 1 & 2 & | & 4 \\ 0 & -5 & -11 & | & -23 \end{pmatrix}$$

So far, this process is identical to our earlier one. Now, however, we only make zero the number (-5) below the first 1 in the second row:

$$\xrightarrow{A_{2,3}(5)} \begin{pmatrix} 1 & 2 & 3 & | & 9 \\ 0 & 1 & 2 & | & 4 \\ 0 & 0 & -1 & | & -3 \end{pmatrix} \xrightarrow{M_3(-1)} \begin{pmatrix} 1 & 2 & 3 & | & 9 \\ 0 & 1 & 2 & | & 4 \\ 0 & 0 & 1 & | & 3 \end{pmatrix}$$

The augmented matrix of the system (and the coefficient matrix) are now in row echelon form and we immediately see that $x_3 = 3$. We then use **back**

substitution to solve for x_2 and then x_1. The second equation reads $x_2 + 2x_3 = 4$. Thus $x_2 + 2(3) = 4$ and $x_2 = -2$. Similarly, from the first equation we obtain $x_1 + 2(-2) + 3(3) = 9$ or $x_1 = 4$. Thus we again obtain the solution $(4, -2, 3)$. The method of solution just employed is called **Gaussian elimination**.

We therefore have two methods for solving our sample systems of equations:

> **i. GAUSS-JORDAN ELIMINATION:**
> Row-reduce the coefficient matrix to reduced row echelon form.
>
> **ii. GAUSSIAN ELIMINATION:**
> Row-reduce the coefficient matrix to row echelon form, solve for the last unknown, and then use back substitution to solve for the other unknowns.

Which method is more useful? It depends. In solving systems of equations on a computer, Gaussian elimination is the preferred method because it involves fewer elementary row operations. We discuss the numerical solution of systems of equations in Sections 8.2 and 8.3. On the other hand, there are times when it is essential to obtain the reduced row echelon form of a matrix (one of these is discussed in Section 2.7). In these cases Gauss-Jordan elimination is the preferred method.

We now turn to the solution of a general system of m equations in n unknowns. Because of our need to do so in Section 2.7, we shall be solving most of the systems by Gauss-Jordan elimination. Keep in mind, however, that Gaussian elimination is sometimes the preferred approach.

The general $m \times n$ system of m linear equations in n unknowns is given by

$$a_{11}x_1 + a_{12}x_2 + a_{13}x_3 + \ldots + a_{1n}x_n = b_1$$
$$a_{21}x_1 + a_{22}x_2 + a_{23}x_3 + \ldots + a_{2n}x_n = b_2$$
$$a_{31}x_1 + a_{32}x_2 + a_{33}x_3 + \ldots + a_{3n}x_n = b_3 \qquad (7)$$
$$\vdots$$
$$a_{m1}x_1 + a_{m2}x_2 + a_{m3}x_3 + \ldots + a_{mn}x_n = b_m$$

In system (7) all the a's and b's are given real numbers. The problem is to find all sets of n numbers, denoted $(x_1, x_2, x_3, \ldots, x_n)$, that satisfy each of the m equations in (7). The number a_{ij} is the coefficient of the variable x_j in the ith equation.

We solve system (7) by writing the system as an augmented matrix and row-reducing the matrix to its reduced row echelon form. We start by dividing the first row by a_{11} (elementary row operation (i)). If $a_{11} = 0$ then we rearrange† the equations so that, with rearrangement, the new $a_{11} \neq 0$. We then use the first equation to eliminate the x_1 term in each of the other equations (using elementary row operation (ii)). Then the new second equation is divided by the new a_{22} term and the new, new second equation is used to eliminate the x_2 terms in all the other equations. The process is continued until one of three situations occurs:

> i. The last nonzero‡ equation reads $x_n = c$ for some constant c. Then there is either a unique solution or an infinite number of solutions to the system.
> ii. The last nonzero equation reads $a'_{ij}x_j + a'_{i,j+1}x_{j+1} + \cdots + a'_{i,j+k}x_n = c$ for some constant c where at least two of the a's are nonzero. That is, the last equation is a linear equation in two or more of the variables. Then there are an infinite number of solutions.
> iii. The last equation reads $0 = c$, where $c \neq 0$. Then there is no solution. In this case the system is called **inconsistent**. In cases (i) and (ii) the system is called **consistent**.

EXAMPLE 7   Solve the system

$$x_1 + 3x_2 - 5x_3 + x_4 = 4$$
$$2x_1 + 5x_2 - 2x_3 + 4x_4 = 6$$

Solution   We write this system as an augmented matrix and row reduce:

$$\begin{pmatrix} 1 & 3 & -5 & 1 & | & 4 \\ 2 & 5 & -2 & 4 & | & 6 \end{pmatrix} \xrightarrow{A_{1,2}(-2)} \begin{pmatrix} 1 & 3 & -5 & 1 & | & 4 \\ 0 & -1 & 8 & 2 & | & -2 \end{pmatrix} \xrightarrow{M_2(-1)}$$

$$\begin{pmatrix} 1 & 3 & -5 & 1 & | & 4 \\ 0 & 1 & -8 & -2 & | & 2 \end{pmatrix} \xrightarrow{A_{2,1}(-3)} \begin{pmatrix} 1 & 0 & 19 & 7 & | & -2 \\ 0 & 1 & -8 & -2 & | & 2 \end{pmatrix}$$

This is as far as we can go. The coefficient matrix is in reduced row echelon form—case (ii) above. There are evidently an infinite number of solutions. The variables x_3 and x_4 can be chosen arbitrarily. Then $x_2 = 2 + 8x_3 + 2x_4$ and $x_1 = -2 - 19x_3 - 7x_4$. All solutions are, therefore, represented by $(-2 - 19x_3 - 7x_4, 2 + 8x_3 + 2x_4, x_3, x_4)$. For example, if $x_3 = 1$ and $x_4 = 2$, we obtain the solution $(-35, 14, 1, 2)$.

As you will see if you do a lot of system solving, the computations can become very messy. It is a good rule of thumb to use a calculator whenever the fractions become unpleasant. It should be noted, however, that if computations are carried out on a computer or calculator, "round-off" errors can be introduced. This problem is discussed in Section 8.1.

---

†To rearrange a system of equations we simply write the same equations in a different order. For example, the first equation can become the fourth equation, the third equation can become the second equation, and so on. This is a sequence of elementary row operations (iii).
‡ The "zero equation" is the equation $0 = 0$.

We close this section with three examples illustrating how a system of linear equations can arise in a practical situation.

EXAMPLE 8   A model that is often used in economics is the **Leontief input-output model**.† Suppose an economic system has n industries. There are two kinds of demands on each industry. First there is the *external* demand from outside the system. If the system is a country, for example, then the external demand could be from another country. Second there is the demand placed on one industry by another industry in the same system. In the United States, for example, there is a demand on the output of the steel industry by the automobile industry.

Let e_i represent the external demand placed on the ith industry. Let a_{ij} represent the internal demand placed on the ith industry by the jth industry. More precisely, a_{ij} represents the number of units of the output of industry i needed to produce 1 unit of the output of industry j. Let x_i represent the output of industry i. Now we assume that the output of each industry is equal to its demand (that is, there is no overproduction). The total demand is equal to the sum of the internal and external demands. To calculate the internal demand on industry 2, for example, we note that $a_{21}x_1$ is the demand on industry 2 made by industry 1. Thus the total internal demand on industry 2 is $a_{21}x_1 + a_{22}x_2 + \cdots + a_{2n}x_n$.

We are led to the following system of equations obtained by equating the total demand with the output of each industry:

$$
\begin{aligned}
a_{11}x_1 + a_{12}x_2 + \cdots + a_{1n}x_n + e_1 &= x_1 \\
a_{21}x_1 + a_{22}x_2 + \cdots + a_{2n}x_n + e_2 &= x_2 \\
&\vdots \\
a_{n1}x_1 + a_{n2}x_2 + \cdots + a_{nn}x_n + e_n &= x_n
\end{aligned}
\tag{8}
$$

Or, rewriting (8) so it looks like system (7), we get

$$
\begin{aligned}
(1-a_{11})x_1 - a_{12}x_2 - \cdots - a_{1n}x_n &= e_1 \\
-a_{21}x_1 + (1-a_{22})x_2 - \cdots - a_{2n}x_n &= e_2 \\
&\vdots \\
-a_{n1}x_1 - a_{n2}x_2 - \cdots + (1-a_{nn})x_n &= e_n
\end{aligned}
\tag{9}
$$

System (9) of n equations in n unknowns is very important in economic analysis.

---

†Named after American economist Wassily W. Leontief. This model was used in his pioneering paper "Qualitative Input and Output Relations in the Economic System of the United States" in *Review of Economic Statistics* 18(1936):105–125. An updated version of this model appears in Leontief's book *Input-Output Analysis* (New York: Oxford University Press, 1966). Leontief won the Nobel Prize in economics in 1973 for his development of input-output analysis.

EXAMPLE 9 In an economic system with three industries, suppose that the external demands are, respectively, 10, 25, and 20. Suppose that $a_{11} = 0.2$, $a_{12} = 0.5$, $a_{13} = 0.15$, $a_{21} = 0.4$, $a_{22} = 0.1$, $a_{23} = 0.3$, $a_{31} = 0.25$, $a_{32} = 0.5$, and $a_{33} = 0.15$. Find the output in each industry such that supply exactly equals demand.

Solution Here $n = 3$, $1 - a_{11} = 0.8$, $1 - a_{22} = 0.9$, and $1 - a_{33} = 0.85$. Then system (9) is

$$0.8x_1 - 0.5x_2 - 0.15x_3 = 10$$
$$-0.4x_1 + 0.9x_2 - 0.3x_3 = 25$$
$$-0.25x_1 - 0.5x_2 + 0.85x_3 = 20$$

Solving this system by using a calculator, we obtain successively (using five-decimal-place accuracy and Gauss-Jordan elimination)

$$\begin{pmatrix} 0.8 & -0.5 & -0.15 & | & 10 \\ -0.4 & 0.9 & -0.3 & | & 25 \\ -0.25 & -0.5 & 0.85 & | & 20 \end{pmatrix} \xrightarrow{M_1(\frac{1}{0.8})} \begin{pmatrix} 1 & -0.625 & -0.1875 & | & 12.5 \\ -0.4 & 0.9 & -0.3 & | & 25 \\ -0.25 & -0.5 & 0.85 & | & 20 \end{pmatrix}$$

$$\xrightarrow[A_{1,3}(0.25)]{A_{1,2}(0.4)} \begin{pmatrix} 1 & -0.625 & -0.1875 & | & 12.5 \\ 0 & 0.65 & -0.375 & | & 30 \\ 0 & -0.65625 & 0.80313 & | & 23.125 \end{pmatrix}$$

$$\xrightarrow{M_2(\frac{1}{0.65})} \begin{pmatrix} 1 & -0.625 & -0.1875 & | & 12.5 \\ 0 & 1 & -0.57692 & | & 46.15385 \\ 0 & -0.65625 & 0.80313 & | & 23.125 \end{pmatrix}$$

$$\xrightarrow[A_{2,3}(0.65625)]{A_{2,1}(0.625)} \begin{pmatrix} 1 & 0 & -0.54808 & | & 41.34616 \\ 0 & 1 & -0.57692 & | & 46.15385 \\ 0 & 0 & 0.42453 & | & 53.41346 \end{pmatrix}$$

$$\xrightarrow{M_3(1/0.42453)} \begin{pmatrix} 1 & 0 & -0.54808 & | & 41.34616 \\ 0 & 1 & -0.57692 & | & 46.15385 \\ 0 & 0 & 1 & | & 125.81787 \end{pmatrix}$$

$$\xrightarrow[A_{3,2}(0.57692)]{A_{3,1}(0.54808)} \begin{pmatrix} 1 & 0 & 0 & | & 110.30442 \\ 0 & 1 & 0 & | & 118.74070 \\ 0 & 0 & 1 & | & 125.81787 \end{pmatrix}$$

We conclude that the outputs needed for supply to equal demand are, approximately, $x_1 = 110$, $x_2 = 119$, and $x_3 = 126$.

EXAMPLE 10 A State Fish and Game Department supplies three types of food to a lake that supports three species of fish. Each fish of Species 1 consumes, each week, an average of 1 unit of Food 1, 1 unit of Food 2, and 2 units of Food 3. Each fish of Species 2 consumes, each week, an average of 3 units of Food 1, 4 units of Food 2, and 5 units of Food 3. For a fish of Species 3, the average weekly consumption is 2 units of Food 1, 1 unit of Food 2, and 5 units of Food 3. Each week 25,000 units of Food 1, 20,000 units of Food 2, and 55,000 units of

1.3 / m EQUATIONS IN n UNKNOWNS: GAUSS-JORDAN AND GAUSSIAN ELIMINATION

Food 3 are supplied to the lake. If we assume that all food is eaten, how many fish of each species can coexist in the lake?

Solution We let x_1, x_2, and x_3 denote the numbers of fish of the three species being supported by the lake environment. Using the information in the problem, we see that x_1 fish of Species 1 consume x_1 units of Food 1, x_2 fish of Species 2 consume $3x_2$ units of Food 1, and x_3 fish of Species 3 consume $2x_3$ units of Food 1. Thus $x_1 + 3x_2 + 2x_3 = 25{,}000 = $ total weekly supply of Food 1. Obtaining a similar equation for each of the other two foods, we are led to the following system:

$$x_1 + 3x_2 + 2x_3 = 25{,}000$$
$$x_1 + 4x_2 + x_3 = 20{,}000$$
$$2x_1 + 5x_2 + 5x_3 = 55{,}000$$

Upon solving, we obtain

$$\begin{pmatrix} 1 & 3 & 2 & | & 25{,}000 \\ 1 & 4 & 1 & | & 20{,}000 \\ 2 & 5 & 5 & | & 55{,}000 \end{pmatrix} \xrightarrow[A_{1,3}(-2)]{A_{1,2}(-1)} \begin{pmatrix} 1 & 3 & 2 & | & 25{,}000 \\ 0 & 1 & -1 & | & -5{,}000 \\ 0 & -1 & 1 & | & 5{,}000 \end{pmatrix} \xrightarrow[A_{2,3}(1)]{A_{2,1}(-3)} \begin{pmatrix} 1 & 0 & 5 & | & 40{,}000 \\ 0 & 1 & -1 & | & -5{,}000 \\ 0 & 0 & 0 & | & 0 \end{pmatrix}$$

Thus, if x_3 is chosen arbitrarily, we have an infinite number of solutions given by $(40{,}000 - 5x_3,\ x_3 - 5{,}000,\ x_3)$. Of course, we must have $x_1 \geq 0$, $x_2 \geq 0$ and $x_3 \geq 0$. Since $x_2 = x_3 - 5{,}000 \geq 0$, we have $x_3 \geq 5{,}000$. This means that $0 \leq x_1 \leq 40{,}000 - 5(5{,}000) = 15{,}000$. Finally, since $40{,}000 - 5x_3 \geq 0$, we see that $x_3 \leq 8{,}000$. This means that the populations that can be supported by the lake with all food consumed are

$$x_1 = 40{,}000 - 5x_3$$
$$x_2 = x_3 - 5{,}000$$
$$5{,}000 \leq x_3 \leq 8{,}000$$

For example, if $x_3 = 6{,}000$, then $x_1 = 10{,}000$ and $x_2 = 1{,}000$.

PROBLEMS 1.3

In Problems 1–20, use Gauss-Jordan and Gaussian elimination to find all solutions, if any, to the given systems.

1. $x_1 - 2x_2 + 3x_3 = 11$
   $4x_1 + x_2 - x_3 = 4$
   $2x_1 - x_2 + 3x_3 = 10$

2. $-2x_1 + x_2 + 6x_3 = 18$
   $5x_1 + 8x_3 = -16$
   $3x_1 + 2x_2 - 10x_3 = -3$

3. $3x_1 + 6x_2 - 6x_3 = 9$
   $2x_1 - 5x_2 + 4x_3 = 6$
   $-x_1 + 16x_2 - 14x_3 = -3$

4. $3x_1 + 6x_2 - 6x_3 = 9$
   $2x_1 - 5x_2 + 4x_3 = 6$
   $5x_1 + 28x_2 - 26x_3 = -8$

5. $x_1 + x_2 - x_3 = 7$
   $4x_1 - x_2 + 5x_3 = 4$
   $2x_1 + 2x_2 - 3x_3 = 0$

6. $x_1 + x_2 - x_3 = 7$
   $4x_1 - x_2 + 5x_3 = 4$
   $6x_1 + x_2 + 3x_3 = 18$

7. $x_1 + x_2 - x_3 = 7$
$4x_1 - x_2 + 5x_3 = 4$
$6x_1 + x_2 + 3x_3 = 20$

8. $x_1 - 2x_2 + 3x_3 = 0$
$4x_1 + x_2 - x_3 = 0$
$2x_1 - x_2 + 3x_3 = 0$

9. $x_1 + x_2 - x_3 = 0$
$4x_1 - x_2 + 5x_3 = 0$
$6x_1 + x_2 + 3x_3 = 0$

10. $2x_2 + 5x_3 = 6$
$x_1 \quad - 2x_3 = 4$
$2x_1 + 4x_2 \quad = -2$

11. $x_1 + 2x_2 - x_3 = 4$
$3x_1 + 4x_2 - 2x_3 = 7$

12. $x_1 + 2x_2 - 4x_3 = 4$
$-2x_1 - 4x_2 + 8x_3 = -8$

13. $x_1 + 2x_2 - 4x_3 = 4$
$-2x_1 - 4x_2 + 8x_3 = -9$

14. $x_1 + 2x_2 - x_3 + x_4 = 7$
$3x_1 + 6x_2 - 3x_3 + 3x_4 = 21$

15. $2x_1 + 6x_2 - 4x_3 + 2x_4 = 4$
$x_1 \quad - x_3 + x_4 = 5$
$-3x_1 + 2x_2 - 2x_3 \quad = -2$

16. $x_1 - 2x_2 + x_3 + x_4 = 2$
$3x_1 \quad + 2x_3 - 2x_4 = -8$
$4x_2 - x_3 - x_4 = 1$
$-x_1 + 6x_2 - 2x_3 \quad = 7$

17. $x_1 - 2x_2 + x_3 + x_4 = 2$
$3x_1 \quad + 2x_3 - 2x_4 = -8$
$4x_2 - x_3 - x_4 = 1$
$5x_1 \quad + 3x_3 - x_4 = -3$

18. $x_1 - 2x_2 + x_3 + x_4 = 2$
$3x_1 \quad + 2x_3 - 2x_4 = -8$
$4x_2 - x_3 - x_4 = 1$
$5x_1 \quad + 3x_3 - x_4 = 0$

19. $x_1 + x_2 = 4$
$2x_1 - 3x_2 = 7$
$3x_1 + 2x_2 = 8$

20. $x_1 + x_2 = 4$
$2x_1 - 3x_2 = 7$
$3x_1 - 2x_2 = 11$

In Problems 21–29 determine whether the given matrix is in row echelon form (but not reduced row echelon form), reduced row echelon form, or neither.

21. $\begin{pmatrix} 1 & 1 & 0 \\ 0 & 1 & 1 \\ 0 & 0 & 1 \end{pmatrix}$

22. $\begin{pmatrix} 2 & 0 & 0 \\ 0 & 1 & 0 \\ 0 & 0 & -1 \end{pmatrix}$

23. $\begin{pmatrix} 1 & 0 & 1 & 0 \\ 0 & 1 & 1 & 0 \\ 0 & 0 & 0 & 0 \end{pmatrix}$

24. $\begin{pmatrix} 1 & 0 & 0 & 0 \\ 0 & 0 & 1 & 0 \\ 0 & 0 & 0 & 1 \end{pmatrix}$

25. $\begin{pmatrix} 0 & 1 & 0 & 0 \\ 1 & 0 & 0 & 0 \\ 0 & 0 & 0 & 0 \end{pmatrix}$

26. $\begin{pmatrix} 1 & 0 & 1 & 2 \\ 0 & 1 & 3 & 4 \end{pmatrix}$

27. $\begin{pmatrix} 1 & 0 \\ 0 & 1 \\ 0 & 0 \end{pmatrix}$

28. $\begin{pmatrix} 1 & 0 & 0 \\ 0 & 0 & 0 \\ 0 & 0 & 1 \end{pmatrix}$

29. $\begin{pmatrix} 1 & 0 & 0 & 4 \\ 0 & 1 & 0 & 5 \\ 0 & 1 & 1 & 6 \end{pmatrix}$

In Problems 30–35 use the elementary row operations to reduce the given matrices to row echelon form and reduced row echelon form.

30. $\begin{pmatrix} 1 & 1 \\ 2 & 3 \end{pmatrix}$

31. $\begin{pmatrix} -1 & 6 \\ 4 & 2 \end{pmatrix}$

32. $\begin{pmatrix} 1 & -1 & 1 \\ 2 & 4 & 3 \\ 5 & 6 & -2 \end{pmatrix}$

33. $\begin{pmatrix} 2 & -4 & 8 \\ 3 & 5 & 8 \\ -6 & 0 & 4 \end{pmatrix}$

34. $\begin{pmatrix} 2 & -4 & -2 \\ 3 & 1 & 6 \end{pmatrix}$

35. $\begin{pmatrix} 2 & -7 \\ 3 & 5 \\ 4 & -3 \end{pmatrix}$

36. In the Leontief input-output model of Example 8, suppose that there are three industries. Suppose further that $e_1 = 10$, $e_2 = 15$, $e_3 = 30$, $a_{11} = \frac{1}{3}$, $a_{12} = \frac{1}{2}$, $a_{13} = \frac{1}{6}$, $a_{21} = \frac{1}{4}$, $a_{22} = \frac{1}{4}$, $a_{23} = \frac{1}{8}$, $a_{31} = \frac{1}{12}$, $a_{32} = \frac{1}{3}$, and $a_{33} = \frac{1}{6}$. Find the output of each industry such that supply exactly equals demand.

37. In Example 10, assume that there are 15,000 units of the first food, 10,000 units of the second, and 35,000 units of the third supplied to the lake each week. Assuming that all three foods are consumed, what populations of the three species can coexist in the lake? Is there a unique solution?

38. A traveler just returned from Europe spent $30 a day for housing in England, $20 a day in France and $20 a day in Spain. For food the traveler spent $20 a day in England, $30 a day in France, and $20 a day in Spain. The traveler spent $10 a day in each country for incidental expenses. The traveler's records of the trip indicate a total of $340 spent for housing, $320 for food, and $140 for incidental expenses while traveling in these countries. Calculate the number of days the traveler spent in each of the countries or show that the records must be incorrect, because the amounts spent are incompatible with each other.

39. An investor remarks to a stockbroker that all her stock holdings are in three companies, Eastern Airlines, Hilton Hotels, and McDonald's, and that 2 days ago the value of her stocks went down $350 but yesterday the value increased by $600. The broker recalls that 2 days ago the price of Eastern Airlines stock dropped by $1 a share, Hilton Hotels dropped $1.50, but the price of McDonald's stock rose by $0.50. The broker also remembers that yesterday the price of Eastern Airlines stock rose $1.50, there was a further drop of $0.50 a share in Hilton Hotels stock, and McDonald's stock rose $1. Show that the broker does not have enough information to calculate the number of shares the investor owns of each company's stock, but that when the investor says that she owns 200 shares of McDonald's stock, the broker can calculate the number of shares of Eastern Airlines and Hilton Hotels.

40. An intelligence agent knows that 60 aircraft, consisting of fighter planes and bombers, are stationed at a certain secret airfield. The agent wishes to determine how many of the 60 are fighter planes and how many are bombers. There is a type of rocket carried by both sorts of planes; the fighter carries six of these rockets, the bomber only two. The agent learns that 250 rockets are required to arm every plane at this airfield. Furthermore, the agent overhears a remark that there are twice as many fighter planes as bombers at the base (that is, the number of fighter planes minus twice the number of bombers equals zero). Calculate the number of fighter planes and bombers at the airfield or show that the agent's information must be incorrect, because it is inconsistent.

41. Consider the system

$$2x_1 - x_2 + 3x_3 = a$$
$$3x_1 + x_2 - 5x_3 = b$$
$$-5x_1 - 5x_2 + 21x_3 = c$$

Show that the system is inconsistent if $c \ne 2a - 3b$.

42. Consider the system

$$2x_1 + 3x_2 - x_3 = a$$
$$x_1 - x_2 + 3x_3 = b$$
$$3x_1 + 7x_2 - 5x_3 = c$$

Find conditions on a, b, and c such that the system is consistent.

*43. Consider the general system of three linear equations in three unknowns:

$$a_{11}x_1 + a_{12}x_2 + a_{13}x_3 = b_1$$
$$a_{21}x_1 + a_{22}x_2 + a_{23}x_3 = b_2$$
$$a_{31}x_1 + a_{32}x_2 + a_{33}x_3 = b_3$$

Find conditions on the coefficients a_{ij} such that the system has a unique solution.

44. Solve the following system using a hand calculator and carrying 5 decimal places of accuracy:

$$2x_2 - x_3 - 4x_4 = 2$$
$$x_1 - x_2 + 5x_3 + 2x_4 = -4$$
$$3x_1 + 3x_2 - 7x_3 - x_4 = 4$$
$$-x_1 - 2x_2 + 3x_3 = -7$$

45. Do the same for the system

$$3.8x_1 + 1.6x_2 + 0.9x_3 = 3.72$$
$$-0.7x_1 + 5.4x_2 + 1.6x_3 = 3.16$$
$$1.5x_1 + 1.1x_2 - 3.2x_3 = 43.78$$

1.4 Homogeneous Systems of Equations

The general $m \times n$ system of linear equations (system (1.3.7), page 15) is called **homogeneous** if all the constants $b_1, b_2, \ldots, b_m$ are zero. That is, the general homogeneous system is given by

$$a_{11}x_1 + a_{12}x_2 + \cdots + a_{1n}x_n = 0$$
$$a_{21}x_1 + a_{22}x_2 + \cdots + a_{2n}x_n = 0$$
$$\vdots \qquad \vdots \qquad \qquad \vdots \qquad \vdots \qquad (1)$$
$$a_{m1}x_1 + a_{m2}x_2 + \cdots + a_{mn}x_n = 0$$

Homogeneous systems arise in a variety of ways. We shall see one of these in the next chapter (in Section 2.6). In this section we shall solve some homogeneous systems—again by the method of Gauss-Jordan elimination.

For the general linear system there are three possibilities: no solution, one solution, or an infinite number of solutions. For the general homogeneous system the situation is simpler. Since $x_1 = x_2 = \cdots = x_n = 0$ is always a solution (called the **trivial solution** or **zero solution**), there are only two possibilities: either the zero solution is the only solution or there are an infinite number of solutions in addition to the zero solution. (Solutions other than the zero solution are called **nontrivial solutions**.

EXAMPLE 1 Solve the homogeneous system

$$2x_1 + 4x_2 + 6x_3 = 0$$
$$4x_1 + 5x_2 + 6x_3 = 0$$
$$3x_1 + x_2 - 2x_3 = 0$$

Solution This is the homogeneous version of the system in Example 1.3.1, page 6.

Reducing successively, we obtain (after dividing the first equation by 2)

$$\begin{pmatrix} 1 & 2 & 3 & | & 0 \\ 4 & 5 & 6 & | & 0 \\ 3 & 1 & -2 & | & 0 \end{pmatrix} \xrightarrow[A_{1,3}(-3)]{A_{1,2}(-4)} \begin{pmatrix} 1 & 2 & 3 & | & 0 \\ 0 & -3 & -6 & | & 0 \\ 0 & -5 & -11 & | & 0 \end{pmatrix} \xrightarrow{M_2(-\frac{1}{3})} \begin{pmatrix} 1 & 2 & 3 & | & 0 \\ 0 & 1 & 2 & | & 0 \\ 0 & -5 & -11 & | & 0 \end{pmatrix}$$

$$\xrightarrow[A_{2,3}(5)]{A_{2,1}(-2)} \begin{pmatrix} 1 & 0 & -1 & | & 0 \\ 0 & 1 & 2 & | & 0 \\ 0 & 0 & -1 & | & 0 \end{pmatrix} \xrightarrow{M_3(-1)} \begin{pmatrix} 1 & 0 & -1 & | & 0 \\ 0 & 1 & 2 & | & 0 \\ 0 & 0 & 1 & | & 0 \end{pmatrix} \xrightarrow[A_{3,2}(-2)]{A_{3,1}(1)} \begin{pmatrix} 1 & 0 & 0 & | & 0 \\ 0 & 1 & 0 & | & 0 \\ 0 & 0 & 1 & | & 0 \end{pmatrix}$$

Thus the system has the unique solution $(0, 0, 0)$. That is, the system has only the trivial solution.

EXAMPLE 2 Solve the homogeneous system

$$x_1 + 2x_2 - x_3 = 0$$
$$3x_1 - 3x_2 + 2x_3 = 0$$
$$-x_1 - 11x_2 + 6x_3 = 0$$

Solution Using Gauss-Jordan elimination we obtain, successively,

$$\begin{pmatrix} 1 & 2 & -1 & | & 0 \\ 3 & -3 & 2 & | & 0 \\ -1 & -11 & 6 & | & 0 \end{pmatrix} \xrightarrow[A_{1,3}(1)]{A_{1,2}(-3)} \begin{pmatrix} 1 & 2 & -1 & | & 0 \\ 0 & -9 & 5 & | & 0 \\ 0 & -9 & 5 & | & 0 \end{pmatrix}$$

$$\xrightarrow{M_2(-\frac{1}{9})} \begin{pmatrix} 1 & 2 & -1 & | & 0 \\ 0 & 1 & -\frac{5}{9} & | & 0 \\ 0 & -9 & 5 & | & 0 \end{pmatrix} \xrightarrow[A_{2,3}(9)]{A_{2,1}(-2)} \begin{pmatrix} 1 & 0 & \frac{1}{9} & | & 0 \\ 0 & 1 & -\frac{5}{9} & | & 0 \\ 0 & 0 & 0 & | & 0 \end{pmatrix}$$

The augmented matrix is now in reduced row echelon form and, evidently, there are an infinite number of solutions given by $(-\frac{1}{9}x_3, \frac{5}{9}x_3, x_3)$. If $x_3 = 0$, for example, we obtain the trivial solution. If $x_3 = 1$ we obtain the solution $(-\frac{1}{9}, \frac{5}{9}, 1)$.

EXAMPLE 3 Solve the system

$$x_1 + x_2 - x_3 = 0$$
$$4x_1 - 2x_2 + 7x_3 = 0$$

(2)

Solution Row-reducing, we obtain

$$\begin{pmatrix} 1 & 1 & -1 & | & 0 \\ 4 & -2 & 7 & | & 0 \end{pmatrix} \xrightarrow{A_{1,2}(-4)} \begin{pmatrix} 1 & 1 & -1 & | & 0 \\ 0 & -6 & 11 & | & 0 \end{pmatrix}$$

$$\xrightarrow{M_2(-\frac{1}{6})} \begin{pmatrix} 1 & 1 & -1 & | & 0 \\ 0 & 1 & -\frac{11}{6} & | & 0 \end{pmatrix} \xrightarrow{A_{2,1}(-1)} \begin{pmatrix} 1 & 0 & \frac{5}{6} & | & 0 \\ 0 & 1 & -\frac{11}{6} & | & 0 \end{pmatrix}$$

Thus there are an infinite number of solutions given by $(-\frac{5}{6}x_3, \frac{11}{6}x_3, x_3)$. This is not surprising since system (2) contains three unknowns and only two equations.

In fact, if there are more unknowns than equations, the homogeneous system (1) will always have an infinite number of solutions. To see this, note that if there were only the trivial solution, then row reduction would lead us to the system

$$x_1 \qquad\qquad = 0$$
$$\quad x_2 \qquad\quad = 0$$
$$\vdots$$
$$x_n = 0$$

and, possibly, additional equations of the form $0 = 0$. But this system has at least as many equations as unknowns. Since row reduction does not change either the number of equations or the number of unknowns, we have a contradiction of our assumption that there were more unknowns than equations. Thus we have Theorem 1.

THEOREM 1 The homogeneous system (1) has an infinite number of solutions if $n > m$.

PROBLEMS 1.4 In Problems 1–13 find all solutions to the homogeneous systems.

1. $2x_1 - x_2 = 0$
   $3x_1 + 4x_2 = 0$

2. $x_1 - 5x_2 = 0$
   $-x_1 + 5x_2 = 0$

3. $x_1 + x_2 - x_3 = 0$
   $2x_1 - 4x_2 + 3x_3 = 0$
   $3x_1 + 7x_2 - x_3 = 0$

4. $x_1 + x_2 - x_3 = 0$
   $2x_1 - 4x_2 + 3x_3 = 0$
   $-x_1 - 7x_2 + 6x_3 = 0$

5. $x_1 + x_2 - x_3 = 0$
   $2x_1 - 4x_2 + 3x_3 = 0$
   $-5x_1 + 13x_2 - 10x_3 = 0$

6. $2x_1 + 3x_2 - x_3 = 0$
   $6x_1 - 5x_2 + 7x_3 = 0$

7. $4x_1 - x_2 = 0$
   $7x_1 + 3x_2 = 0$
   $-8x_1 + 6x_2 = 0$

8. $x_1 - x_2 + 7x_3 - x_4 = 0$
   $2x_1 + 3x_2 - 8x_3 + x_4 = 0$

9. $x_1 - 2x_2 + x_3 + x_4 = 0$
   $3x_1 \qquad + 2x_3 - 2x_4 = 0$
   $\qquad 4x_2 - x_3 - x_4 = 0$
   $5x_1 \qquad + 3x_3 - x_4 = 0$

10. $-2x_1 \qquad\qquad + 7x_4 = 0$
    $x_1 + 2x_2 - x_3 + 4x_4 = 0$
    $3x_1 \qquad - x_3 + 5x_4 = 0$
    $4x_1 + 2x_2 + 3x_3 \qquad = 0$

11. $2x_1 - x_2 = 0$
    $3x_1 + 5x_2 = 0$
    $7x_1 - 3x_2 = 0$
    $-2x_1 + 3x_2 = 0$

12. $x_1 - 3x_2 = 0$
    $-2x_1 + 6x_2 = 0$
    $4x_1 - 12x_2 = 0$

13. $x_1 + x_2 - x_3 = 0$
    $4x_1 - x_2 + 5x_3 = 0$
    $-2x_1 + x_2 - 2x_3 = 0$
    $3x_1 + 2x_2 - 6x_3 = 0$

14. Show that the homogeneous system

$$a_{11}x_1 + a_{12}x_2 = 0$$
$$a_{21}x_1 + a_{22}x_2 = 0$$

has an infinite number of solutions if and only if the determinant of the system is zero.

15. Consider the system
$$2x_1 - 3x_2 + 5x_3 = 0$$
$$-x_1 + 7x_2 - x_3 = 0$$
$$4x_1 - 11x_2 + kx_3 = 0$$

For what value of k will the system have nontrivial solutions?

*16. Consider the 3×3 homogeneous system
$$a_{11}x_1 + a_{12}x_2 + a_{13}x_3 = 0$$
$$a_{21}x_1 + a_{22}x_2 + a_{23}x_3 = 0$$
$$a_{31}x_1 + a_{32}x_2 + a_{33}x_3 = 0$$

Find conditions on the coefficients a_{ij} such that the zero solution is the only solution.

Review Exercises for Chapter 1

In Exercises 1–14, find all solutions (if any) to the given systems.

1. $3x_1 + 6x_2 = 9$
$-2x_1 + 3x_2 = 4$

2. $3x_1 + 6x_2 = 9$
$2x_1 + 4x_2 = 6$

3. $3x_1 - 6x_2 = 9$
$-2x_1 + 4x_2 = 6$

4. $x_1 + x_2 + x_3 = 2$
$2x_1 - x_2 + 2x_3 = 4$
$-3x_1 + 2x_2 + 3x_3 = 8$

5. $x_1 + x_2 + x_3 = 0$
$2x_1 - x_2 + 2x_3 = 0$
$-3x_1 + 2x_2 + 3x_3 = 0$

6. $x_1 + x_2 + x_3 = 2$
$2x_1 - x_2 + 2x_3 = 4$
$-x_1 + 4x_2 + x_3 = 2$

7. $x_1 + x_2 + x_3 = 2$
$2x_1 - x_2 + 2x_3 = 4$
$-x_1 + 4x_2 + x_3 = 3$

8. $x_1 + x_2 + x_3 = 0$
$2x_1 - x_2 + 2x_3 = 0$
$-x_1 + 4x_2 + x_3 = 0$

9. $2x_1 + x_2 - 3x_3 = 0$
$4x_1 - x_2 + x_3 = 0$

10. $x_1 + x_2 = 0$
$2x_1 + x_2 = 0$
$3x_1 + x_2 = 0$

11. $x_1 + x_2 = 1$
$2x_1 + x_2 = 3$
$3x_1 + x_2 = 4$

12. $x_1 + x_2 + x_3 + x_4 = 4$
$2x_1 - 3x_2 - x_3 + 4x_4 = 7$
$-2x_1 + 4x_2 + x_3 - 2x_4 = 1$
$5x_1 - x_2 + 2x_3 + x_4 = -1$

13. $x_1 + x_2 + x_3 + x_4 = 0$
$2x_1 - 3x_2 - x_3 + 4x_4 = 0$
$-2x_1 + 4x_2 + x_3 - 2x_4 = 0$
$5x_1 - x_2 + 2x_3 + x_4 = 0$

14. $x_1 + x_2 + x_3 + x_4 = 0$
$2x_1 - 3x_2 - x_3 + 4x_4 = 0$
$-2x_1 + 4x_2 + x_3 - 2x_4 = 0$

15. Find the distance from the point $(3, -2)$ to the line $x - 2y = 6$.

In Exercises 16–20 determine whether the given matrix is in row echelon form (but not reduced row echelon form), reduced row echelon form, or neither.

16. $\begin{pmatrix} 1 & 0 & 0 & 0 \\ 0 & 1 & 0 & 2 \\ 0 & 0 & 1 & 3 \end{pmatrix}$

17. $\begin{pmatrix} 1 & 8 & 1 & 0 \\ 0 & 1 & 5 & -7 \\ 0 & 0 & 1 & 4 \end{pmatrix}$

18. $\begin{pmatrix} 1 & 0 \\ 0 & 3 \\ 0 & 0 \end{pmatrix}$

19. $\begin{pmatrix} 1 & 0 & 2 & 0 \\ 0 & 1 & 3 & 0 \end{pmatrix}$ **20.** $\begin{pmatrix} 1 & 1 & 1 & 1 \\ 0 & 1 & 1 & 1 \end{pmatrix}$

In Exercises 21 and 22, reduce the matrix to row echelon form and reduced row echelon form.

21. $\begin{pmatrix} 2 & 8 & -2 \\ 1 & 0 & -6 \end{pmatrix}$ **22.** $\begin{pmatrix} 1 & -1 & 2 & 4 \\ -1 & 2 & 0 & 3 \\ 2 & 3 & -1 & 1 \end{pmatrix}$

2 Vectors and Matrices

2.1 Vectors

William Rowan Hamilton
(Granger Collection)

The study of vectors began essentially with the work of the great Irish mathematician Sir William Rowan Hamilton (1805–1865). Hamilton was a genius who, by the age of twelve, had mastered not only the languages of continental Europe but also Greek, Latin, Sanskrit, Hebrew, Chinese, Persian, Arabic, Malay, Hindi, Bengali, and several others as well. In his twenties, Hamilton turned to science, and his treatises on mechanics and optics provide the basis for much of modern physics.

In his thirties this remarkable man began his research in mathematics. His desire to find a way to represent certain objects in the plane and in space led to the discovery of what he called "quaternions." This notion led to the development of what we now call *vectors*. Throughout Hamilton's life, and for the remainder of the nineteenth century, there was considerable debate over the usefulness of quaternions and vectors. At the end of the century, the great British physicist Lord Kelvin wrote that quaternions, "although beautifully ingenious, have been an unmixed evil to those who have touched them in any way [and] vectors... have never been of the slightest use to any creature."

But Kelvin was wrong. Today nearly all branches of classical and modern physics are represented by means of the language of vectors. Vectors are also used with increasing frequency in the social and biological sciences.[†]

On page 2 we described the solution to a system of two equations in two unknowns to be a pair of numbers written (x_1, x_2). In Example 1.3.1 on page 8 we wrote the solution of the system of three equations in three unknowns as the triple of numbers $(2, -1, 4)$. Both (x_1, x_2) and $(2, -1, 4)$ are **vectors**.

n-COMPONENT ROW VECTOR

We define an **n-component row vector** to be an **ordered** set of n numbers written as

$$(x_1, x_2, \ldots, x_n) \tag{1}$$

[†] For interesting discussions of the development of modern vector analysis, consult the book by M. J. Crowe, *A History of Vector Analysis* (Notre Dame: University of Notre Dame Press, 1967) or Morris Kline's excellent book *Mathematical Thought from Ancient to Modern Times* (New York: Oxford University Press, 1972), chap. 32.

n-COMPONENT COLUMN VECTOR

An *n*-component column vector is an **ordered** set of *n* numbers written as

$$\begin{pmatrix} x_1 \\ x_2 \\ \vdots \\ x_n \end{pmatrix} \qquad (2)$$

In (1) or (2), x_1 is called the **first component** of the vector, x_2 is the **second component**, and so on. In general, x_k is called the **kth component** of the vector.

For simplicity, we shall often refer to an *n*-component row vector as a **row vector** or an *n*-**vector**. Similarly, we shall use the term **column vector** (or *n*-vector) to denote an *n*-component column vector. Any vector whose entries are all zero is called a **zero vector**.

EXAMPLE 1

The following are vectors:

i. $(3, 6)$ is a row vector (or a 2-vector).

ii. $\begin{pmatrix} 2 \\ -1 \\ 5 \end{pmatrix}$ is a column vector (or a 3-vector).

iii. $(2, -1, 0, 4)$ is a row vector (or a 4-vector).

iv. $\begin{pmatrix} 0 \\ 0 \\ 0 \\ 0 \\ 0 \end{pmatrix}$ is a column vector and a zero vector.

Warning. The word "ordered" in the definition of a vector is essential. Two vectors with the same components written in different orders are *not* the same. Thus, for example, the row vectors $(1, 2)$ and $(2, 1)$ are not equal.

For the remainder of this text we shall denote vectors with boldface lowercase letters like **u, v, a, b, c,** and so on. A zero vector is denoted **0**.

Vectors arise in a great number of ways. Suppose that the buyer for a manufacturing plant must order different quantities of steel, aluminum, oil, and paper. He can keep track of the quantities to be ordered with a single vector. The vector $\begin{pmatrix} 10 \\ 30 \\ 15 \\ 60 \end{pmatrix}$ indicates that he would order 10 units of steel, 30 units of aluminum, and so on.

Remark. We see here why the order in which the components of a vector are written is important. It is clear that the vectors $\begin{pmatrix} 30 \\ 15 \\ 60 \\ 10 \end{pmatrix}$ and $\begin{pmatrix} 10 \\ 30 \\ 15 \\ 60 \end{pmatrix}$ mean very different things to the buyer.

It is time to describe some properties of vectors. Since it would be repetitive to do so first for row vectors and then for column vectors, we shall give all definitions in terms of column vectors. Similar definitions hold for row vectors.

The components of all the vectors in this text are either real or complex numbers.† We use the symbol $\mathbb{R}^n$ to denote the set of all n-vectors $\begin{pmatrix} a_1 \\ a_2 \\ \vdots \\ a_n \end{pmatrix}$, where each a_i is a real number. Similarly, we use the symbol $\mathbb{C}^n$ to denote the set of all n-vectors $\begin{pmatrix} c_1 \\ c_2 \\ \vdots \\ c_n \end{pmatrix}$, where each c_i is a complex number. In Chapter 4 we shall discuss the sets $\mathbb{R}^2$ (vectors in the plane) and $\mathbb{R}^3$ (vectors in space). In Chapter 5 we shall examine arbitrary sets of vectors.

DEFINITION 1 **EQUALITY OF VECTORS** Two column (or row) vectors **a** and **b** are **equal** if and only if‡ they have the same number of components and their corresponding components are equal. In symbols, the vectors $\mathbf{a} = \begin{pmatrix} a_1 \\ a_2 \\ \vdots \\ a_n \end{pmatrix}$ and $\mathbf{b} = \begin{pmatrix} b_1 \\ b_2 \\ \vdots \\ b_n \end{pmatrix}$ are equal if and only if $a_1 = b_1, a_2 = b_2, \ldots, a_n = b_n$.

DEFINITION 2 **ADDITION OF VECTORS** Let $\mathbf{a} = \begin{pmatrix} a_1 \\ a_2 \\ \vdots \\ a_n \end{pmatrix}$ and $\mathbf{b} = \begin{pmatrix} b_1 \\ b_2 \\ \vdots \\ b_n \end{pmatrix}$ be n-vectors. Then the sum of **a** and **b** is defined by

$$\mathbf{a} + \mathbf{b} = \begin{pmatrix} a_1 + b_1 \\ a_2 + b_2 \\ \vdots \\ a_n + b_n \end{pmatrix} \tag{3}$$

---

† A complex number is a number of the form $a + ib$, where a and b are real numbers and $i = \sqrt{-1}$. A description of complex numbers is given in Appendix 2. We shall not encounter complex vectors again until Chapter 5; they will be especially useful in Chapter 7. Therefore, unless otherwise stated, we assume, for the time being, that all vectors have real components.

‡ The term "if and only if" applies to two statements. "Statement A if and only if statement B" means that statements A and B are equivalent. That is, if statement A is true then statement B is true and if statement B is true then statement A is true. Put another way, it means that you cannot have one without the other.

EXAMPLE 2
$$\begin{pmatrix} 1 \\ 2 \\ 4 \end{pmatrix} + \begin{pmatrix} -6 \\ 7 \\ 5 \end{pmatrix} = \begin{pmatrix} -5 \\ 9 \\ 9 \end{pmatrix}$$

EXAMPLE 3
$$\begin{pmatrix} 2 \\ -1 \end{pmatrix} + \begin{pmatrix} -2 \\ 1 \end{pmatrix} = \begin{pmatrix} 0 \\ 0 \end{pmatrix}$$

Warning. It is essential that **a** and **b** have the same number of components. For example, the sum $\begin{pmatrix} 2 \\ 3 \end{pmatrix} + \begin{pmatrix} 1 \\ 2 \\ 3 \end{pmatrix}$ is not defined since 2-vectors and 3-vectors are different kinds of objects and cannot be added together. Moreover, it is not possible to add a row and a column vector together. For example, the sum $\begin{pmatrix} 1 \\ 2 \end{pmatrix} + (3, 5)$ is *not* defined.

When dealing with vectors, we shall refer to numbers as **scalars** (which may be real or complex depending on whether the vectors in question are real or complex).†

DEFINITION 3 **SCALAR MULTIPLICATION OF VECTORS** Let $\mathbf{a} = \begin{pmatrix} a_1 \\ a_2 \\ \vdots \\ a_n \end{pmatrix}$ be a vector and α a scalar. Then the product $\alpha \mathbf{a}$ is given by

$$\alpha \mathbf{a} = \begin{pmatrix} \alpha a_1 \\ \alpha a_2 \\ \vdots \\ \alpha a_n \end{pmatrix} \qquad (4)$$

That is, to multiply a vector by a scalar, we simply multiply each component of the vector by the scalar.

---

† *Historical Note:* The term "scalar" originated with Hamilton. His definition of the quaternion included what he called a "real part" and an "imaginary part." In his paper "On Quaternions, or on a New System of Imaginaries in Algebra," in *Philosophical Magazine*, 3rd series, 25(1844): 26–27 he wrote: "The algebraically *real* part may receive... all values contained on the one *scale* of progression of numbers from negative to positive infinity; we shall call it therefore the *scalar part*, or simply the *scalar* of the quaternion...." In the same paper, Hamilton went on to define the imaginary part of his quaternion as the *vector* part. Although this was not the first usage of the word "vector," it was the first time it was used in the context of the definitions in this section. It is fair to say that the paper from which the preceding quotation was taken marks the beginning of modern vector analysis.

EXAMPLE 4

$$3\begin{pmatrix} 2 \\ -1 \\ 4 \end{pmatrix} = \begin{pmatrix} 6 \\ -3 \\ 12 \end{pmatrix}$$

Note. Putting Definition 1 and Definition 2 together, we can define the difference of two vectors by

$$\mathbf{a} - \mathbf{b} = \mathbf{a} + (-1)\mathbf{b} \tag{5}$$

This means that if $\mathbf{a} = \begin{pmatrix} a_1 \\ a_2 \\ \vdots \\ a_n \end{pmatrix}$ and $\mathbf{b} = \begin{pmatrix} b_1 \\ b_2 \\ \vdots \\ b_n \end{pmatrix}$, then $\mathbf{a} - \mathbf{b} = \begin{pmatrix} a_1 - b_1 \\ a_2 - b_2 \\ \vdots \\ a_n - b_n \end{pmatrix}$.

EXAMPLE 5

Let $\mathbf{a} = \begin{pmatrix} 4 \\ 6 \\ 1 \\ 3 \end{pmatrix}$ and $\mathbf{b} = \begin{pmatrix} -2 \\ 4 \\ -3 \\ 0 \end{pmatrix}$. Calculate $2\mathbf{a} - 3\mathbf{b}$.

Solution
$$2\mathbf{a} - 3\mathbf{b} = 2\begin{pmatrix} 4 \\ 6 \\ 1 \\ 3 \end{pmatrix} + (-3)\begin{pmatrix} -2 \\ 4 \\ -3 \\ 0 \end{pmatrix} = \begin{pmatrix} 8 \\ 12 \\ 2 \\ 6 \end{pmatrix} + \begin{pmatrix} 6 \\ -12 \\ 9 \\ 0 \end{pmatrix} = \begin{pmatrix} 14 \\ 0 \\ 11 \\ 6 \end{pmatrix}$$

Once we know how to add vectors and multiply them by scalars, we can prove a number of facts relating these operations. Several of these facts are given in Theorem 1 below. We prove parts (*ii*) and (*iii*) and leave the remaining parts as exercises (see Problems 21–23).

THEOREM 1 Let $\mathbf{a}$, $\mathbf{b}$, and $\mathbf{c}$ be n-vectors and let α and β be scalars. Then:

  i. $\mathbf{a} + \mathbf{0} = \mathbf{a}$
  ii. $0\mathbf{a} = \mathbf{0}$ (Note that the zero on the left is the number zero, whereas the zero on the right is a zero vector.)
  iii. $\mathbf{a} + \mathbf{b} = \mathbf{b} + \mathbf{a}$ (commutative law)
  iv. $(\mathbf{a} + \mathbf{b}) + \mathbf{c} = \mathbf{a} + (\mathbf{b} + \mathbf{c})$ (associative law)
  v. $\alpha(\mathbf{a} + \mathbf{b}) = \alpha\mathbf{a} + \alpha\mathbf{b}$ (distributive law for scalar multiplication)
  vi. $(\alpha + \beta)\mathbf{a} = \alpha\mathbf{a} + \beta\mathbf{a}$
  vii. $(\alpha\beta)\mathbf{a} = \alpha(\beta\mathbf{a})$

Proof of (ii) and (iii)

ii. If $\mathbf{a} = \begin{pmatrix} a_1 \\ a_2 \\ \vdots \\ a_n \end{pmatrix}$, then $0\mathbf{a} = 0\begin{pmatrix} a_1 \\ a_2 \\ \vdots \\ a_n \end{pmatrix} = \begin{pmatrix} 0 \cdot a_1 \\ 0 \cdot a_2 \\ \vdots \\ 0 \cdot a_n \end{pmatrix} = \begin{pmatrix} 0 \\ 0 \\ \vdots \\ 0 \end{pmatrix} = \mathbf{0}$.

iii. Let $\mathbf{b} = \begin{pmatrix} b_1 \\ b_2 \\ \vdots \\ b_n \end{pmatrix}$. Then $\mathbf{a} + \mathbf{b} = \begin{pmatrix} a_1 + b_1 \\ a_2 + b_2 \\ \vdots \\ a_n + b_n \end{pmatrix} = \begin{pmatrix} b_1 + a_1 \\ b_2 + a_2 \\ \vdots \\ b_n + a_n \end{pmatrix} = \mathbf{b} + \mathbf{a}.$

Here we used the fact that for any two numbers x and y, $x + y = y + x$ and $0 \cdot x = 0$. ∎

Note. In (ii), the zero on the left is the scalar zero (that is, the real number 0) and the zero on the right is the zero vector. These two things are different.

EXAMPLE 6

To illustrate the associative law, we note that

$$\left[\begin{pmatrix} 3 \\ 1 \\ 2 \end{pmatrix} + \begin{pmatrix} -2 \\ 4 \\ -1 \end{pmatrix}\right] + \begin{pmatrix} 6 \\ -3 \\ 5 \end{pmatrix} = \begin{pmatrix} 1 \\ 5 \\ 1 \end{pmatrix} + \begin{pmatrix} 6 \\ -3 \\ 5 \end{pmatrix} = \begin{pmatrix} 7 \\ 2 \\ 6 \end{pmatrix}$$

while

$$\begin{pmatrix} 3 \\ 1 \\ 2 \end{pmatrix} + \left[\begin{pmatrix} -2 \\ 4 \\ -1 \end{pmatrix} + \begin{pmatrix} 6 \\ -3 \\ 5 \end{pmatrix}\right] = \begin{pmatrix} 3 \\ 1 \\ 2 \end{pmatrix} + \begin{pmatrix} 4 \\ 1 \\ 4 \end{pmatrix} = \begin{pmatrix} 7 \\ 2 \\ 6 \end{pmatrix}$$

Example 6 illustrates the importance of the associative law of vector addition, since if we wish to add together three or more vectors, we can only do so by adding them together two at a time. The associative law tells us that we can do this in two different ways and still come up with the same answer. If this were not the case, the sum of three or more vectors would be more difficult to define since we would have to specify whether we wanted $(\mathbf{a} + \mathbf{b}) + \mathbf{c}$ or $\mathbf{a} + (\mathbf{b} + \mathbf{c})$ to be the definition of the sum $\mathbf{a} + \mathbf{b} + \mathbf{c}$.

PROBLEMS 2.1

In Problems 1–10 perform the indicated computation with $\mathbf{a} = \begin{pmatrix} -3 \\ 1 \\ 4 \end{pmatrix}$, $\mathbf{b} = \begin{pmatrix} 5 \\ -4 \\ 7 \end{pmatrix}$, and $\mathbf{c} = \begin{pmatrix} 2 \\ 0 \\ -2 \end{pmatrix}$.

1. $\mathbf{a} + \mathbf{b}$
2. $3\mathbf{b}$
3. $-2\mathbf{c}$
4. $\mathbf{b} + 3\mathbf{c}$
5. $2\mathbf{a} - 5\mathbf{b}$
6. $-3\mathbf{b} + 2\mathbf{c}$
7. $0\mathbf{c}$
8. $\mathbf{a} + \mathbf{b} + \mathbf{c}$
9. $3\mathbf{a} - 2\mathbf{b} + 4\mathbf{c}$
10. $3\mathbf{b} - 7\mathbf{c} + 2\mathbf{a}$

In Problems 11–20 perform the indicated computation with $\mathbf{a} = (3, -1, 4, 2)$, $\mathbf{b} = (6, 0, -1, 4)$, and $\mathbf{c} = (-2, 3, 1, 5)$. Of course, it is first necessary to extend the definitions in this section to row vectors.

11. $\mathbf{a} + \mathbf{c}$
12. $\mathbf{b} - \mathbf{a}$
13. $4\mathbf{c}$
14. $-2\mathbf{b}$
15. $2\mathbf{a} - \mathbf{c}$
16. $4\mathbf{b} - 7\mathbf{a}$

17. $\mathbf{a}+\mathbf{b}+\mathbf{c}$
18. $\mathbf{c}-\mathbf{b}+2\mathbf{a}$
19. $3\mathbf{a}-2\mathbf{b}+4\mathbf{c}$
20. $\alpha\mathbf{a}+\beta\mathbf{b}+\gamma\mathbf{c}$

21. Let $\mathbf{a} = \begin{pmatrix} a_1 \\ a_2 \\ \vdots \\ a_n \end{pmatrix}$ and let $\mathbf{0}$ denote the n-component zero column vector. Use Definitions 2 and 3 to show that $\mathbf{a}+\mathbf{0}=\mathbf{a}$ and $0\mathbf{a}=\mathbf{0}$.

22. Let $\mathbf{a} = \begin{pmatrix} a_1 \\ a_2 \\ \vdots \\ a_n \end{pmatrix}$, $\mathbf{b} = \begin{pmatrix} b_1 \\ b_2 \\ \vdots \\ b_n \end{pmatrix}$, and $\mathbf{c} = \begin{pmatrix} c_1 \\ c_2 \\ \vdots \\ c_n \end{pmatrix}$. Compute $(\mathbf{a}+\mathbf{b})+\mathbf{c}$ and $\mathbf{a}+(\mathbf{b}+\mathbf{c})$ and show that they are equal.

23. Let $\mathbf{a}$ and $\mathbf{b}$ be as in Problem 22 and let α and β be scalars. Compute $\alpha(\mathbf{a}+\mathbf{b})$ and $\alpha\mathbf{a}+\alpha\mathbf{b}$ and show that they are equal. Similarly, compute $(\alpha+\beta)\mathbf{a}$ and $\alpha\mathbf{a}+\beta\mathbf{a}$ and show that they are equal. Finally, show that $(\alpha\beta)\mathbf{a}=\alpha(\beta\mathbf{a})$.

24. Find numbers α, β, and γ such that $(2,-1,4)+(\alpha,\beta,\gamma)=\mathbf{0}$.

25. In the manufacture of a certain product, four raw materials are needed. The vector $\mathbf{d} = \begin{pmatrix} d_1 \\ d_2 \\ d_3 \\ d_4 \end{pmatrix}$ represents a given factory's demand for each of the four raw materials to produce 1 unit of its product. If $\mathbf{d}_1$ is the demand vector for factory 1 and $\mathbf{d}_2$ is the demand vector for factory 2, what is represented by the vectors $\mathbf{d}_1+\mathbf{d}_2$ and $2\mathbf{d}_1$?

26. Let $\mathbf{a} = \begin{pmatrix} 1 \\ 3 \\ 2 \end{pmatrix}$, $\mathbf{b} = \begin{pmatrix} -2 \\ 4 \\ 1 \end{pmatrix}$, and $\mathbf{c} = \begin{pmatrix} 0 \\ 1 \\ 4 \end{pmatrix}$. Find a vector $\mathbf{v}$ such that $2\mathbf{a}-\mathbf{b}+3\mathbf{v}=4\mathbf{c}$.

27. With $\mathbf{a}$, $\mathbf{b}$, and $\mathbf{c}$ as in Problem 26, find a vector $\mathbf{w}$ such that $\mathbf{a}-\mathbf{b}+\mathbf{c}-\mathbf{w}=\mathbf{0}$.

2.2 The Scalar Product of Two Vectors

In Section 2.1 we saw how two vectors could be added and multiplied by scalars. There are several ways that two vectors can be multiplied together. Two of these ways are discussed in this text. In this section we define a product of two vectors the result of which is a scalar. In Section 4.4 we show how the product of two vectors can yield a vector.

DEFINITION 1   **SCALAR PRODUCT**   Let $\mathbf{a} = \begin{pmatrix} a_1 \\ a_2 \\ \vdots \\ a_n \end{pmatrix}$ and $\mathbf{b} = \begin{pmatrix} b_1 \\ b_2 \\ \vdots \\ b_n \end{pmatrix}$ be two n-vectors. Then the **scalar product** of $\mathbf{a}$ and $\mathbf{b}$, denoted $\mathbf{a} \cdot \mathbf{b}$, is given by

$$\mathbf{a} \cdot \mathbf{b} = a_1 b_1 + a_2 b_2 + \cdots + a_n b_n \tag{1}$$

Because of the notation in (1), the scalar product of two vectors is often called the **dot product** of the vectors. Note that the scalar product of two n-vectors is a scalar (that is, a number).

Warning. When taking the scalar product of **a** and **b**, it is necessary that **a** and **b** have the same number of components.

We shall often be taking the scalar product of a row vector and column vector. In this case we have

$$(a_1, a_2, \ldots, a_n) \cdot \begin{pmatrix} b_1 \\ b_2 \\ \vdots \\ b_n \end{pmatrix} = a_1 b_1 + a_2 b_2 + \cdots + a_n b_n \qquad (2)$$

EXAMPLE 1 Let $\mathbf{a} = \begin{pmatrix} 1 \\ -2 \\ 3 \end{pmatrix}$ and $\mathbf{b} = \begin{pmatrix} 3 \\ -2 \\ 4 \end{pmatrix}$. Calculate $\mathbf{a} \cdot \mathbf{b}$.

Solution $\mathbf{a} \cdot \mathbf{b} = (1)(3) + (-2)(-2) + (3)(4) = 3 + 4 + 12 = 19$

EXAMPLE 2 Let $\mathbf{a} = (2, -3, 4, -6)$ and $\mathbf{b} = \begin{pmatrix} 1 \\ 2 \\ 0 \\ 3 \end{pmatrix}$. Compute $\mathbf{a} \cdot \mathbf{b}$.

Solution Here $\mathbf{a} \cdot \mathbf{b} = (2)(1) + (-3)(2) + (4)(0) + (-6)(3) = 2 - 6 + 0 - 18 = -22$.

EXAMPLE 3 Suppose that a manufacturer produces four items. The demand for the items is given by the demand vector $\mathbf{d} = (30, 20, 40, 10)$. The price per unit that he receives for the items is given by the price vector $\mathbf{p} = (\$20, \$15, \$18, \$40)$. If he meets his demand, how much money will he receive?

Solution His demand for the first item is 30 and he receives \$20 for each of the first item sold. He therefore receives $(30)(20) = \$600$ from the sale of the first item. By continuing this reasoning we see that the total cash received is given by $\mathbf{d} \cdot \mathbf{p}$. Thus income received $= \mathbf{d} \cdot \mathbf{p} = (30)(20) + (20)(15) + (40)(18) + (10)(40) = 600 + 300 + 720 + 400 = \$2020$.

The next result follows directly from the definition of the scalar product (see Problem 22).

THEOREM 1 Let **a**, **b**, and **c** be n-vectors and let α and β be scalars. Then:

i. $\mathbf{a} \cdot \mathbf{0} = 0$
ii. $\mathbf{a} \cdot \mathbf{b} = \mathbf{b} \cdot \mathbf{a}$ (commutative law for scalar product)
iii. $\mathbf{a} \cdot (\mathbf{b} + \mathbf{c}) = \mathbf{a} \cdot \mathbf{b} + \mathbf{a} \cdot \mathbf{c}$ (distributive law for scalar product)
iv. $(\alpha \mathbf{a}) \cdot \mathbf{b} = \alpha (\mathbf{a} \cdot \mathbf{b})$

Note that there is *no* associative law for the scalar product. The expression $(\mathbf{a} \cdot \mathbf{b}) \cdot \mathbf{c} = \mathbf{a} \cdot (\mathbf{b} \cdot \mathbf{c})$ does not make sense because neither side of the equation is defined. For the left side, this follows from the fact that $\mathbf{a} \cdot \mathbf{b}$ is a scalar and the scalar product of the scalar $\mathbf{a} \cdot \mathbf{b}$ and the vector **c** is not defined.

EXAMPLE 4 Let $\mathbf{a} = \begin{pmatrix} 1 \\ 2 \\ 4 \end{pmatrix}$, $\mathbf{b} = \begin{pmatrix} -3 \\ 1 \\ 7 \end{pmatrix}$, and $\mathbf{c} = \begin{pmatrix} 2 \\ 5 \\ -1 \end{pmatrix}$. Then $\mathbf{b} + \mathbf{c} = \begin{pmatrix} -1 \\ 6 \\ 6 \end{pmatrix}$. We verify that the distributive law for scalar products (part (*iii*) of Theorem 1) holds for these vectors. We have

$$\mathbf{a} \cdot \mathbf{b} = -3 + 2 + 28 = 27 \quad \text{and} \quad \mathbf{a} \cdot \mathbf{c} = 2 + 10 - 4 = 8$$

Next we compute

$$\mathbf{a} \cdot (\mathbf{b} + \mathbf{c}) = -1 + 12 + 24 = 35$$

and

$$\mathbf{a} \cdot \mathbf{b} + \mathbf{a} \cdot \mathbf{c} = 27 + 8 = 35$$

Finally, we observe that

$$\mathbf{a} \cdot (\mathbf{b} + \mathbf{c}) = \mathbf{a} \cdot \mathbf{b} + \mathbf{a} \cdot \mathbf{c}$$

PROBLEMS 2.2 In Problems 1–7 calculate the scalar product of the two vectors.

1. $\begin{pmatrix} 2 \\ 3 \\ -5 \end{pmatrix}$; $\begin{pmatrix} 3 \\ 0 \\ 4 \end{pmatrix}$

2. $(1, 2, -1, 0)$; $(3, -7, 4, -2)$

3. $\begin{pmatrix} 5 \\ 7 \end{pmatrix}$; $\begin{pmatrix} 3 \\ -2 \end{pmatrix}$

4. $(8, 3, 1)$; $(7, -4, 3)$

5. (a, b); (c, d)

6. $\begin{pmatrix} x \\ y \\ z \end{pmatrix}$; $\begin{pmatrix} y \\ z \\ x \end{pmatrix}$

7. $(-1, -3, 4, 5)$; $(-1, -3, 4, 5)$

8. Let **a** be an n-vector. Show that $\mathbf{a} \cdot \mathbf{a} \geq 0$.

9. Find conditions on a vector **a** such that $\mathbf{a} \cdot \mathbf{a} = 0$.

In Problems 10–14 perform the indicated computation with $\mathbf{a} = \begin{pmatrix} 1 \\ -2 \\ 4 \end{pmatrix}$, $\mathbf{b} = \begin{pmatrix} 0 \\ -3 \\ -7 \end{pmatrix}$, and $\mathbf{c} = \begin{pmatrix} 4 \\ -1 \\ 5 \end{pmatrix}$.

10. $(2\mathbf{a}) \cdot (3\mathbf{b})$
11. $\mathbf{a} \cdot (\mathbf{b} + \mathbf{c})$
12. $\mathbf{c} \cdot (\mathbf{a} - \mathbf{b})$
13. $(2\mathbf{b}) \cdot (3\mathbf{c} - 5\mathbf{a})$
14. $(\mathbf{a} - \mathbf{c}) \cdot (3\mathbf{b} - 4\mathbf{a})$

ORTHOGONAL VECTORS

Two vectors $\mathbf{a}$ and $\mathbf{b}$ are said to be **orthogonal** if $\mathbf{a} \cdot \mathbf{b} = 0$. In Problems 15–19 determine which pairs of vectors are orthogonal.†

15. $\begin{pmatrix} 2 \\ -3 \end{pmatrix}$; $\begin{pmatrix} 3 \\ 2 \end{pmatrix}$
16. $\begin{pmatrix} 2 \\ -3 \end{pmatrix}$; $\begin{pmatrix} -3 \\ 2 \end{pmatrix}$
17. $\begin{pmatrix} 1 \\ 4 \\ -7 \end{pmatrix}$; $\begin{pmatrix} 2 \\ 3 \\ 2 \end{pmatrix}$

18. $(1, 0, 1, 0)$; $(0, 1, 0, 1)$
19. $\begin{pmatrix} a \\ 0 \\ b \\ 0 \\ c \end{pmatrix}$; $\begin{pmatrix} 0 \\ d \\ 0 \\ e \\ 0 \end{pmatrix}$

20. Determine a number α such that $(1, -2, 3, 5)$ is orthogonal to $(-4, \alpha, 6, -1)$.
21. Determine all numbers α and β such that the vectors $\begin{pmatrix} 1 \\ -\alpha \\ 2 \\ 3 \end{pmatrix}$ and $\begin{pmatrix} 4 \\ 5 \\ -2\beta \\ 7 \end{pmatrix}$ are orthogonal.
22. Using the definition of the scalar product, prove Theorem 1.
23. A manufacturer of custom-designed jewelry has orders for two rings, three pairs of earrings, five pins, and one necklace. The manufacturer estimates that it takes 1 hour of labor to make a ring, $1\frac{1}{2}$ hours to make a pair of earrings, $\frac{1}{2}$ hour for each pin, and 2 hours to make a necklace.
    (a) Express the manufacturer's orders as a row vector.
    (b) Express the hourly requirements for the various types of jewelry as a column vector.
    (c) Use the scalar product to calculate the total number of hours it will require to complete all the orders.
24. A tourist returned from a European trip with the following foreign currency: 1000 Austrian schillings, 20 British pounds, 100 French francs, 5000 Italian lire, and 50 German marks. In American money, a schilling was worth $0.055, the pound $1.80, the franc $0.20, the lira $0.001, and the mark $0.40.
    (a) Express the quantity of each currency by means of a row vector.
    (b) Express the value of each currency in American money by means of a column vector.
    (c) Use the scalar product to compute how much the tourist's foreign currency was worth in American money.

---

† We shall be dealing extensively with orthogonal vectors in Chapters 4 and 5.

2.3 Matrices

MATRIX

An $m \times n$ **matrix**† A is a rectangular array of mn numbers arranged in m rows and n columns:‡

$$A = \begin{pmatrix} a_{11} & a_{12} & \cdots & a_{1j} & \cdots & a_{1n} \\ a_{21} & a_{22} & \cdots & a_{2j} & \cdots & a_{2n} \\ \vdots & \vdots & & \vdots & & \vdots \\ a_{i1} & a_{i2} & \cdots & a_{ij} & \cdots & a_{in} \\ \vdots & \vdots & & \vdots & & \vdots \\ a_{m1} & a_{m2} & \cdots & a_{mj} & \cdots & a_{mn} \end{pmatrix} \qquad (1)$$

The ijth component of A, denoted a_{ij}, is the number appearing in the ith row and jth column of A. We will sometimes write the matrix A as $A = (a_{ij})$.
Usually, matrices will be denoted by capital letters.

If A is an $m \times n$ matrix with $m = n$, then A is called a **square matrix**. An $m \times n$ matrix with all components equal to zero is called the $m \times n$ **zero matrix**.

An $m \times n$ matrix is said to have the **size** $m \times n$. Two matrices $A = (a_{ij})$ and $B = (b_{ij})$ are **equal** if (i) they have the same size and (ii) corresponding components are equal.

EXAMPLE 1

Five matrices of different sizes are given below:

i. $A = \begin{pmatrix} 1 & 3 \\ 4 & 2 \end{pmatrix}$, 2×2 (square)   ii. $A = \begin{pmatrix} -1 & 3 \\ 4 & 0 \\ 1 & -2 \end{pmatrix}$, 3×2

iii. $\begin{pmatrix} -1 & 4 & 1 \\ 3 & 0 & 2 \end{pmatrix}$, 2×3 iv. $\begin{pmatrix} 1 & 6 & -2 \\ 3 & 1 & 4 \\ 2 & -6 & 5 \end{pmatrix}$, 3×3 (square)

v. $\begin{pmatrix} 0 & 0 & 0 & 0 \\ 0 & 0 & 0 & 0 \end{pmatrix}$, 2×4 zero matrix

Each vector is a special kind of matrix. Thus, for example, the n-component row vector $(a_1, a_2, \ldots, a_n)$ is a $1 \times n$ matrix whereas the n-component column vector $\begin{pmatrix} a_1 \\ a_2 \\ \vdots \\ a_n \end{pmatrix}$ is an $n \times 1$ matrix.

---

† *Historical note:* The term "matrix" was first used in 1850 by the British mathematician James Joseph Sylvester (1814–1897) to distinguish matrices from determinants (which we shall discuss in Chapter 3). In fact, the term "matrix" was intended to mean "mother of determinants."

‡ As with vectors, we shall always assume, unless stated otherwise, that the numbers in a matrix are real.

Matrices, like vectors, arise in a great number of practical situations. For example, we saw in Section 2.1 how the vector $\begin{pmatrix} 10 \\ 30 \\ 15 \\ 60 \end{pmatrix}$ could represent order quantities for four different products used by one manufacturer. Suppose that there were five different plants. Then the 4×5 matrix

$$Q = \begin{pmatrix} 10 & 20 & 15 & 16 & 25 \\ 30 & 10 & 20 & 25 & 22 \\ 15 & 22 & 18 & 20 & 13 \\ 60 & 40 & 50 & 35 & 45 \end{pmatrix}$$

could represent the orders for the four products in each of the five plants. We can see, for example, that plant 4 orders 25 units of the second product while plant 2 orders 40 units of the fourth product.

Matrices, like vectors, can be added and multiplied by scalars.†

DEFINITION 1 **ADDITION OF MATRICES** Let $A = (a_{ij})$ and $B = (b_{ij})$ be two $m \times n$ matrices. Then the sum of A and B is the $m \times n$ matrix $A + B$ given by

$$A + B = (a_{ij} + b_{ij}) = \begin{pmatrix} a_{11}+b_{11} & a_{12}+b_{12} & \cdots & a_{1n}+b_{1n} \\ a_{21}+b_{21} & a_{22}+b_{22} & \cdots & a_{2n}+b_{2n} \\ \vdots & \vdots & & \vdots \\ a_{m1}+b_{m1} & a_{m2}+b_{m2} & \cdots & a_{mn}+b_{mn} \end{pmatrix}$$

That is, $A + B$ is the $m \times n$ matrix obtained by adding the corresponding components of A and B.

Arthur Cayley
(Library of Congress)

---

† The algebra of matrices, that is, the rules by which matrices can be added and multiplied, was developed by the English mathematician Arthur Cayley (1821–1895) in 1857. Matrices arose with Cayley in connection with linear transformations of the type

$$x' = ax + by,$$
$$y' = cx + dy,$$

where a, b, c, d are real numbers, and which may be thought of as mapping the point (x, y) into the point (x', y'). Clearly, the above transformation is completely determined by the four coefficients a, b, c, d, and so the transformation can be symbolized by the square array

$$\begin{pmatrix} a & b \\ c & d \end{pmatrix},$$

which we have called a (*square*) *matrix*. We shall discuss linear transformations in Chapter 6.

Warning. The sum of two matrices is defined only when both matrices have the same size. Thus, for example, it is not possible to add together the matrices $\begin{pmatrix} 1 & 2 & 3 \\ 4 & 5 & 6 \end{pmatrix}$ and $\begin{pmatrix} -1 & 0 \\ 2 & -5 \\ 4 & 7 \end{pmatrix}$.

EXAMPLE 2

$$\begin{pmatrix} 2 & 4 & -6 & 7 \\ 1 & 3 & 2 & 1 \\ -4 & 3 & -5 & 5 \end{pmatrix} + \begin{pmatrix} 0 & 1 & 6 & -2 \\ 2 & 3 & 4 & 3 \\ -2 & 1 & 4 & 4 \end{pmatrix} = \begin{pmatrix} 2 & 5 & 0 & 5 \\ 3 & 6 & 6 & 4 \\ -6 & 4 & -1 & 9 \end{pmatrix}$$

DEFINITION 2

MULTIPLICATION OF A MATRIX BY A SCALAR If $A = (a_{ij})$ is an $m \times n$ matrix and if α is a scalar, then the $m \times n$ matrix αA is given by

$$\alpha A = (\alpha a_{ij}) = \begin{pmatrix} \alpha a_{11} & \alpha a_{12} & \cdots & \alpha a_{1n} \\ \alpha a_{21} & \alpha a_{22} & \cdots & \alpha a_{2n} \\ \vdots & \vdots & & \vdots \\ \alpha a_{m1} & \alpha a_{m2} & \cdots & \alpha a_{mn} \end{pmatrix} \quad (3)$$

In other words, $\alpha A = (\alpha a_{ij})$ is the matrix obtained by multiplying each component of A by α.

EXAMPLE 3

Let $A = \begin{pmatrix} 1 & -3 & 4 & 2 \\ 3 & 1 & 4 & 6 \\ -2 & 3 & 5 & 7 \end{pmatrix}$. Then $2A = \begin{pmatrix} 2 & -6 & 8 & 4 \\ 6 & 2 & 8 & 12 \\ -4 & 6 & 10 & 14 \end{pmatrix}$,

$-3A = \begin{pmatrix} -3 & 9 & -12 & -6 \\ -9 & -3 & -12 & -18 \\ 6 & -9 & -15 & -21 \end{pmatrix}$, and $0A = \begin{pmatrix} 0 & 0 & 0 & 0 \\ 0 & 0 & 0 & 0 \\ 0 & 0 & 0 & 0 \end{pmatrix}$.

EXAMPLE 4

Let $A = \begin{pmatrix} 1 & 2 & 4 \\ -7 & 3 & -2 \end{pmatrix}$ and $B = \begin{pmatrix} 4 & 0 & 5 \\ 1 & -3 & 6 \end{pmatrix}$. Calculate $-2A + 3B$.

Solution $-2A + 3B = (-2)\begin{pmatrix} 1 & 2 & 4 \\ -7 & 3 & -2 \end{pmatrix} + (3)\begin{pmatrix} 4 & 0 & 5 \\ 1 & -3 & 6 \end{pmatrix} =$

$\begin{pmatrix} -2 & -4 & -8 \\ 14 & -6 & 4 \end{pmatrix} + \begin{pmatrix} 12 & 0 & 15 \\ 3 & -9 & 18 \end{pmatrix} = \begin{pmatrix} 10 & -4 & 7 \\ 17 & -15 & 22 \end{pmatrix}$

The next theorem is similar to Theorem 2.1.1 on page 31. Its proof is left as an exercise (see Problems 21–24).

THEOREM 1 Let A, B, and C be $m \times n$ matrices and let α be a scalar. Then:

i. $A + 0 = A$
ii. $0A = 0$ (Note that the zero on the left is the number zero and the zero on the right is a zero matrix.)
iii. $A + B = B + A$ (commutative law for matrix addition)
iv. $(A + B) + C = A + (B + C)$ (associative law for matrix addition)
v. $\alpha(A + B) = \alpha A + \alpha B$ (distributive law for scalar multiplication)
vi. $1A = A$

Note. The zero in part (i) of the theorem is the $m \times n$ zero matrix. In part (ii) the zero on the left is a scalar while the zero on the right is the $m \times n$ zero matrix.

EXAMPLE 5 To illustrate the associative law we note that

$$\left[\begin{pmatrix} 1 & 4 & -2 \\ 3 & -1 & 0 \end{pmatrix} + \begin{pmatrix} 2 & -2 & 3 \\ 1 & -1 & 5 \end{pmatrix}\right] + \begin{pmatrix} 3 & -1 & 2 \\ 0 & 1 & 4 \end{pmatrix} = \begin{pmatrix} 3 & 2 & 1 \\ 4 & -2 & 5 \end{pmatrix}$$

$$+ \begin{pmatrix} 3 & -1 & 2 \\ 0 & 1 & 4 \end{pmatrix} = \begin{pmatrix} 6 & 1 & 3 \\ 4 & -1 & 9 \end{pmatrix}$$

Similarly,

$$\begin{pmatrix} 1 & 4 & -2 \\ 3 & -1 & 0 \end{pmatrix} + \left[\begin{pmatrix} 2 & -2 & 3 \\ 1 & -1 & 5 \end{pmatrix} + \begin{pmatrix} 3 & -1 & 2 \\ 0 & 1 & 4 \end{pmatrix}\right] = \begin{pmatrix} 1 & 4 & -2 \\ 3 & -1 & 0 \end{pmatrix}$$

$$+ \begin{pmatrix} 5 & -3 & 5 \\ 1 & 0 & 9 \end{pmatrix} = \begin{pmatrix} 6 & 1 & 3 \\ 4 & -1 & 9 \end{pmatrix}$$

As with vectors, the associative law for matrix addition enables us to define the sum of three or more matrices.

PROBLEMS 2.3 In Problems 1–12 perform the indicated computation with $A = \begin{pmatrix} 1 & 3 \\ 2 & 5 \\ -1 & 2 \end{pmatrix}$, $B = \begin{pmatrix} -2 & 0 \\ 1 & 4 \\ -7 & 5 \end{pmatrix}$, and $C = \begin{pmatrix} -1 & 1 \\ 4 & 6 \\ -7 & 3 \end{pmatrix}$.

1. $3A$
2. $A + B$
3. $A - C$
4. $2C - 5A$
5. $0B$ (0 is the scalar zero)
6. $-7A + 3B$
7. $A + B + C$
8. $C - A - B$
9. $2A - 3B + 4C$
10. $7C - B + 2A$

11. Find a matrix D such that $2A + B - D$ is the 3×2 zero matrix.
12. Find a matrix E such that $A + 2B - 3C + E$ is the 3×2 zero matrix.

In Problems 13–20 perform the indicated computation with $A = \begin{pmatrix} 1 & -1 & 2 \\ 3 & 4 & 5 \\ 0 & 1 & -1 \end{pmatrix}$,

$B = \begin{pmatrix} 0 & 2 & 1 \\ 3 & 0 & 5 \\ 7 & -6 & 0 \end{pmatrix}$, and $C = \begin{pmatrix} 0 & 0 & 2 \\ 3 & 1 & 0 \\ 0 & -2 & 4 \end{pmatrix}$.

13. $A - 2B$
14. $3A - C$
15. $A + B + C$
16. $2A - B + 2C$
17. $C - A - B$
18. $4C - 2B + 3A$
19. Find a matrix D such that $A + B + C + D$ is the 3×3 zero matrix.
20. Find a matrix E such that $3C - 2B + 8A - 4E$ is the 3×3 zero matrix.
21. Let $A = (a_{ij})$ be an $m \times n$ matrix and let $\bar{0}$ denote the $m \times n$ zero matrix. Use Definitions 1 and 2 to show that $0A = \bar{0}$ and $\bar{0} + A = A$. Similarly, show that $1A = A$.
22. Let $A = (a_{ij})$ and $B = (b_{ij})$ be $m \times n$ matrices. Compute $A + B$ and $B + A$ and show that they are equal.
23. If α is a scalar and A and B are as in Problem 22, compute $\alpha(A + B)$ and $\alpha A + \alpha B$ and show that they are equal.
24. If $A = (a_{ij})$, $B = (b_{ij})$, and $C = (c_{ij})$ are $m \times n$ matrices, compute $(A + B) + C$ and $A + (B + C)$ and show that they are equal.
25. Consider the "graph" joining the four points in the figure. Construct a 4×4 matrix having the property that $a_{ij} = 0$ if point i is not connected (joined by a line) to point j and $a_{ij} = 1$ if point i is connected to point j.

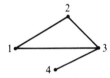

26. Do the same (this time constructing a 5×5 matrix) for the accompanying graph.

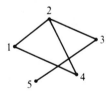

2.4 Matrix Products

In this section we see how two matrices can be multiplied together. Quite obviously, we could define the product of two $m \times n$ matrices $A = (a_{ij})$ and $B = (b_{ij})$ to be the $m \times n$ matrix whose ijth component is $a_{ij}b_{ij}$. However, for just about all the important applications involving matrices, another kind of product is needed. It comes as the generalization of the scalar product.

DEFINITION 1   **PRODUCT OF TWO MATRICES**   Let $A = (a_{ij})$ be an $m \times n$ matrix whose ith row is denoted $\mathbf{a}_i$. Let $B = (b_{ij})$ be an $n \times p$ matrix whose jth column is denoted $\mathbf{b}_j$. Then the product of A and B is an $m \times p$ matrix $C = (c_{ij})$, where

$$c_{ij} = \mathbf{a}_i \cdot \mathbf{b}_j \tag{1}$$

That is, the ijth element of AB is the scalar product of the ith row of A ($\mathbf{a}_i$) and the jth column of B ($\mathbf{b}_j$). If we write this out, we obtain

$$c_{ij} = a_{i1}b_{1j} + a_{i2}b_{2j} + \cdots + a_{in}b_{nj} \tag{2}$$

Warning. Two matrices can be multiplied together only if the number of columns of the first is equal to the number of rows of the second. Otherwise the vectors $\mathbf{a}_i$ and $\mathbf{b}_j$ will have different numbers of components and the scalar product in Equation (1) will not be defined.

EXAMPLE 1   If $A = \begin{pmatrix} 1 & 3 \\ -2 & 4 \end{pmatrix}$ and $B = \begin{pmatrix} 3 & -2 \\ 5 & 6 \end{pmatrix}$, calculate AB and BA.

Solution   Let $C = (c_{ij}) = AB$. Then $c_{11} = \mathbf{a}_1 \cdot \mathbf{b}_1 = (1 \;\; 3) \cdot \begin{pmatrix} 3 \\ 5 \end{pmatrix} = 3 + 15 = 18$; $c_{12} = \mathbf{a}_1 \cdot \mathbf{b}_2 = (1 \;\; 3) \cdot \begin{pmatrix} -2 \\ 6 \end{pmatrix} = -2 + 18 = 16$; $c_{21} = (-2 \;\; 4) \cdot \begin{pmatrix} 3 \\ 5 \end{pmatrix} = -6 + 20 = 14$; and $c_{22} = (-2 \;\; 4) \cdot \begin{pmatrix} -2 \\ 6 \end{pmatrix} = 4 + 24 = 28$. Thus $C = AB = \begin{pmatrix} 18 & 16 \\ 14 & 28 \end{pmatrix}$. Similarly, leaving out the intermediate steps, we see that

$$C' = BA = \begin{pmatrix} 3 & -2 \\ 5 & 6 \end{pmatrix}\begin{pmatrix} 1 & 3 \\ -2 & 4 \end{pmatrix} = \begin{pmatrix} 3+4 & 9-8 \\ 5-12 & 15+24 \end{pmatrix} = \begin{pmatrix} 7 & 1 \\ -7 & 39 \end{pmatrix}$$

Remark. Example 1 illustrates an important fact: *Matrix products do not, in general, commute.* That is, $AB \neq BA$ in general. It sometimes happens that $AB = BA$, but this will be the exception, not the rule. In fact, as the next example illustrates, it may occur that AB is defined while BA is not. Thus we must be careful of *order* when multiplying two matrices together.

EXAMPLE 2   Let $A = \begin{pmatrix} 2 & 0 & -3 \\ 4 & 1 & 5 \end{pmatrix}$ and $B = \begin{pmatrix} 7 & -1 & 4 & 7 \\ 2 & 5 & 0 & -4 \\ -3 & 1 & 2 & 3 \end{pmatrix}$. Calculate AB.

Solution We first note that A is a 2×3 matrix and B is a 3×4 matrix. Hence the number of columns of A equals the number of rows of B. The product AB is therefore defined and is a 2×4 matrix. Let $AB = C = (c_{ij})$. Then

$$c_{11} = (2 \ 0 \ -3) \cdot \begin{pmatrix} 7 \\ 2 \\ -3 \end{pmatrix} = 23 \qquad c_{12} = (2 \ 0 \ -3) \cdot \begin{pmatrix} -1 \\ 5 \\ 1 \end{pmatrix} = -5$$

$$c_{13} = (2 \ 0 \ -3) \cdot \begin{pmatrix} 4 \\ 0 \\ 2 \end{pmatrix} = 2 \qquad c_{14} = (2 \ 0 \ -3) \cdot \begin{pmatrix} 7 \\ -4 \\ 3 \end{pmatrix} = 5$$

$$c_{21} = (4 \ 1 \ 5) \cdot \begin{pmatrix} 7 \\ 2 \\ -3 \end{pmatrix} = 15 \qquad c_{22} = (4 \ 1 \ 5) \cdot \begin{pmatrix} -1 \\ 5 \\ 1 \end{pmatrix} = 6$$

$$c_{23} = (4 \ 1 \ 5) \cdot \begin{pmatrix} 4 \\ 0 \\ 2 \end{pmatrix} = 26 \qquad c_{24} = (4 \ 1 \ 5) \cdot \begin{pmatrix} 7 \\ -4 \\ 3 \end{pmatrix} = 39$$

Hence $AB = \begin{pmatrix} 23 & -5 & 2 & 5 \\ 15 & 6 & 26 & 39 \end{pmatrix}$. This completes the problem. Note that the product BA is *not* defined since the number of columns of B (four) is not equal to the number of rows of A (two).

EXAMPLE 3 **Direct and Indirect Contact with a Contagious Disease.** In this example we show how matrix multiplication can be used to model the spread of a contagious disease. Suppose that four individuals have contracted such a disease. This group has contacts with six people in a second group. We can represent these contacts, called *direct contacts*, by a 4×6 matrix. An example of such a matrix is given below:

DIRECT CONTACT MATRIX: First and second groups

$$A = \begin{pmatrix} 0 & 1 & 0 & 0 & 1 & 0 \\ 1 & 0 & 0 & 1 & 0 & 1 \\ 0 & 0 & 0 & 1 & 1 & 0 \\ 1 & 0 & 0 & 0 & 0 & 1 \end{pmatrix}$$

Here we set $a_{ij} = 1$ if the ith person in the first group has made contact with the jth person in the second group. For example, the 1 in the 2,4 position means that the second person in the first (infected) group has been in contact with the fourth person in the second group. Now suppose that a third group of five people has had a variety of direct contact with individuals of the second group. We can also represent this by a matrix.

DIRECT CONTACT MATRIX: Second and third groups

$$B = \begin{pmatrix} 0 & 0 & 1 & 0 & 1 \\ 0 & 0 & 0 & 1 & 0 \\ 0 & 1 & 0 & 0 & 0 \\ 1 & 0 & 0 & 0 & 1 \\ 0 & 0 & 0 & 1 & 0 \\ 0 & 0 & 1 & 0 & 0 \end{pmatrix}$$

Note that $b_{64} = 0$, which means that the sixth person in the second group has had no contact with the fourth person in the third group.

The *indirect* or *second-order* contacts between the individuals in the first and third groups is represented by the 4×5 matrix $C = AB$. To see this, observe that a person in group 3 can be infected from someone in group 2 who, in turn, has been infected by someone in group 1. For example, since $a_{24} = 1$ and $b_{45} = 1$, we see that, indirectly, the fifth person in group 3 has contact (through the fourth person in group 2) with the second person in group 1. The total number of indirect contacts between the second person in group 1 and the fifth person in group 3 is given by

$$c_{25} = a_{21}b_{15} + a_{22}b_{25} + a_{23}b_{35} + a_{24}b_{45} + a_{25}b_{55} + a_{26}b_{26}$$
$$= 1\cdot 1 + 0\cdot 0 + 0\cdot 0 + 1\cdot 1 + 0\cdot 0 + 1\cdot 0 = 2$$

We now compute.

INDIRECT CONTACT MATRIX: First and third groups

$$C = AB = \begin{pmatrix} 0 & 0 & 0 & 2 & 0 \\ 1 & 0 & 2 & 0 & 2 \\ 1 & 0 & 0 & 1 & 1 \\ 0 & 0 & 2 & 0 & 1 \end{pmatrix}$$

We observe that only the second person in Group 3 has no indirect contacts with the disease. The fifth person in this group has $2 + 1 + 1 = 4$ indirect contacts.

We have seen that for matrix multiplication the commutative law does not hold. The next theorem shows that the associative law does hold.

THEOREM 1

ASSOCIATIVE LAW FOR MATRIX MULTIPLICATION Let $A = (a_{ij})$ be an $n \times m$ matrix, $B = (b_{ij})$ an $m \times p$ matrix, and $C = (c_{ij})$ a $p \times q$ matrix. Then the **associative law**

$$A(BC) = (AB)C \quad (3)$$

holds and ABC, defined by either side of (3), is an $n \times q$ matrix.

The proof of this theorem is not difficult, but it is somewhat tedious. It is best given using the summation notation. (If this is unfamiliar take a look at Problems 36–56.) For that reason let us defer it until the end of the section.

EXAMPLE 4 Verify the associative law for $A = \begin{pmatrix} 1 & -3 \\ 0 & 2 \end{pmatrix}$, $B = \begin{pmatrix} 2 & -1 & 4 \\ 3 & 1 & 5 \end{pmatrix}$, and $C = \begin{pmatrix} 0 & -2 & 1 \\ 4 & 3 & 2 \\ -5 & 0 & 6 \end{pmatrix}$.

Solution We first note that A is 2×2, B is 2×3, and C is 3×3. Hence all products used in the statement of the associative law are defined and the resulting product will be a 2×3 matrix. We then calculate

$$AB = \begin{pmatrix} 1 & -3 \\ 0 & 2 \end{pmatrix}\begin{pmatrix} 2 & -1 & 4 \\ 3 & 1 & 5 \end{pmatrix} = \begin{pmatrix} -7 & -4 & -11 \\ 6 & 2 & 10 \end{pmatrix}$$

$$(AB)C = \begin{pmatrix} -7 & -4 & -11 \\ 6 & 2 & 10 \end{pmatrix}\begin{pmatrix} 0 & -2 & 1 \\ 4 & 3 & 2 \\ -5 & 0 & 6 \end{pmatrix} = \begin{pmatrix} 39 & 2 & -81 \\ -42 & -6 & 70 \end{pmatrix}$$

Similarly,

$$BC = \begin{pmatrix} 2 & -1 & 4 \\ 3 & 1 & 5 \end{pmatrix}\begin{pmatrix} 0 & -2 & 1 \\ 4 & 3 & 2 \\ -5 & 0 & 6 \end{pmatrix} = \begin{pmatrix} -24 & -7 & 24 \\ -21 & -3 & 35 \end{pmatrix}$$

$$A(BC) = \begin{pmatrix} 1 & -3 \\ 0 & 2 \end{pmatrix}\begin{pmatrix} -24 & -7 & 24 \\ -21 & -3 & 35 \end{pmatrix} = \begin{pmatrix} 39 & 2 & -81 \\ -42 & -6 & 70 \end{pmatrix}$$

Thus $(AB)C = A(BC)$.

From now on we shall write the product of three matrices simply as ABC. We can do this because $(AB)C = A(BC)$; thus we get the same answer no matter how the multiplication is carried out (provided that we do not commute any of the matrices).

The associative law can be extended to longer products. For example, suppose that AB, BC, and CD are defined. Then

$$ABCD = A(B(CD)) = ((AB)C)D = A(BC)D = (AB)(CD) \qquad (4)$$

There are two distributive laws for matrix multiplication.

THEOREM 2 **DISTRIBUTIVE LAWS FOR MATRIX MULTIPLICATION** If all the following sums and products are defined, then

$$A(B+C) = AB + AC \tag{5}$$

and
$$(A+B)C = AC + BC \tag{6}$$

Proofs of Theorems 1 and 2

Associative Laws. Since A is $n \times m$ and B is $m \times p$, AB is $n \times p$. Thus $(AB)C = (n \times p) \times (p \times q)$ is an $n \times q$ matrix. Similarly BC is $m \times q$ and $A(BC)$ is $n \times q$ so that $(AB)C$ and $A(BC)$ are both of the same size. We must show that the ijth component of $(AB)C$ equals the ijth component of $A(BC)$. Define $D = (d_{ij}) = AB$. Then $d_{ij} = \sum_{k=1}^{m} a_{ik} b_{kj}$. The ijth component of $(AB)C = DC$ is $\sum_{l=1}^{p} d_{il} c_{lj} = \sum_{l=1}^{p} \left(\sum_{k=1}^{m} a_{ik} b_{kl} \right) c_{lj} = \sum_{k=1}^{m} \sum_{l=1}^{p} a_{ik} b_{kl} c_{lj}$. Next we define $E = (e_{ij}) = BC$. Then $e_{ij} = \sum_{l=1}^{p} b_{il} c_{lj}$ and the ijth component of $A(BC) = AE$ is $\sum_{k=1}^{m} a_{ik} e_{kj} = \sum_{k=1}^{m} \sum_{l=1}^{p} a_{ik} b_{kl} c_{lj}$. Thus the ijth component of $(AB)C$ is equal to the ijth component of $A(BC)$. This proves the associative law. ∎

Distributive Laws. We prove the first distributive law (Equation 5). The proof of the second one (Equation 6) is virtually identical and is therefore omitted. Let A be $n \times m$ and let B and C be $m \times p$. Then the kjth component of $B+C$ is $b_{kj} + c_{kj}$ and the ijth component of $A(B+C)$ is $\sum_{k=1}^{m} a_{ik}(b_{kj} + c_{kj}) = \sum_{k=1}^{m} a_{ik} b_{kj} + \sum_{k=1}^{m} a_{ik} c_{kj} = ij$th component of AB plus the ijth component of AC and this proves Equation (5). ∎

PROBLEMS 2.4

In Problems 1–15 perform the indicated computation.

1. $\begin{pmatrix} 2 & 3 \\ -1 & 2 \end{pmatrix} \begin{pmatrix} 4 & 1 \\ 0 & 6 \end{pmatrix}$
2. $\begin{pmatrix} 3 & -2 \\ 1 & 4 \end{pmatrix} \begin{pmatrix} -5 & 6 \\ 1 & 3 \end{pmatrix}$
3. $\begin{pmatrix} 1 & -1 \\ 1 & 1 \end{pmatrix} \begin{pmatrix} -1 & 0 \\ 2 & 3 \end{pmatrix}$

4. $\begin{pmatrix} -5 & 6 \\ 1 & 3 \end{pmatrix} \begin{pmatrix} 3 & -2 \\ 1 & 4 \end{pmatrix}$
5. $\begin{pmatrix} -4 & 5 & 1 \\ 0 & 4 & 2 \end{pmatrix} \begin{pmatrix} 3 & -1 & 1 \\ 5 & 6 & 4 \\ 0 & 1 & 2 \end{pmatrix}$
6. $\begin{pmatrix} 7 & 1 & 4 \\ 2 & -3 & 5 \end{pmatrix} \begin{pmatrix} 1 & 6 \\ 0 & 4 \\ -2 & 3 \end{pmatrix}$

7. $\begin{pmatrix} 1 & 6 \\ 0 & 4 \\ -2 & 3 \end{pmatrix} \begin{pmatrix} 7 & 1 & 4 \\ 2 & -3 & 5 \end{pmatrix}$
8. $\begin{pmatrix} 1 & 4 & -2 \\ 3 & 0 & 4 \end{pmatrix} \begin{pmatrix} 0 & 1 \\ 2 & 3 \end{pmatrix}$
9. $\begin{pmatrix} 1 & 4 & 6 \\ -2 & 3 & 5 \\ 1 & 0 & 4 \end{pmatrix} \begin{pmatrix} 2 & -3 & 5 \\ 1 & 0 & 6 \\ 2 & 3 & 1 \end{pmatrix}$

10. $\begin{pmatrix} 2 & -3 & 5 \\ 1 & 0 & 6 \\ 2 & 3 & 1 \end{pmatrix} \begin{pmatrix} 1 & 4 & 6 \\ -2 & 3 & 5 \\ 1 & 0 & 4 \end{pmatrix}$
11. $(1 \quad 4 \quad 0 \quad 2) \begin{pmatrix} 3 & -6 \\ 2 & 4 \\ 1 & 0 \\ -2 & 3 \end{pmatrix}$

12. $\begin{pmatrix} 3 & 2 & 1 & -2 \\ -6 & 4 & 0 & 3 \end{pmatrix} \begin{pmatrix} 1 \\ 4 \\ 0 \\ 2 \end{pmatrix}$

13. $\begin{pmatrix} 3 & -2 & 1 \\ 4 & 0 & 6 \\ 5 & 1 & 9 \end{pmatrix} \begin{pmatrix} 1 & 0 & 0 \\ 0 & 1 & 0 \\ 0 & 0 & 1 \end{pmatrix}$

14. $\begin{pmatrix} 1 & 0 & 0 \\ 0 & 1 & 0 \\ 0 & 0 & 1 \end{pmatrix} \begin{pmatrix} 3 & -2 & 1 \\ 4 & 0 & 6 \\ 5 & 1 & 9 \end{pmatrix}$

15. $\begin{pmatrix} a & b & c \\ d & e & f \\ g & h & j \end{pmatrix} \begin{pmatrix} 1 & 0 & 0 \\ 0 & 1 & 0 \\ 0 & 0 & 1 \end{pmatrix}$, where $a, b, c, d, e, f, g, h, j$ are real numbers.

16. Find a matrix $A = \begin{pmatrix} a & b \\ c & d \end{pmatrix}$ such that $A \begin{pmatrix} 2 & 3 \\ 1 & 2 \end{pmatrix} = \begin{pmatrix} 1 & 0 \\ 0 & 1 \end{pmatrix}$.

*17. Let a_{11}, a_{12}, a_{21}, and a_{22} be given real numbers such that $a_{11}a_{22} - a_{12}a_{21} \neq 0$. Find numbers b_{11}, b_{12}, b_{21}, and b_{22} such that $\begin{pmatrix} a_{11} & a_{12} \\ a_{21} & a_{22} \end{pmatrix} \begin{pmatrix} b_{11} & b_{12} \\ b_{21} & b_{22} \end{pmatrix} = \begin{pmatrix} 1 & 0 \\ 0 & 1 \end{pmatrix}$.

18. Verify the associative law for multiplication for the matrices $A = \begin{pmatrix} 2 & -1 & 4 \\ 1 & 0 & 6 \end{pmatrix}$, $B = \begin{pmatrix} 1 & 0 & 1 \\ 2 & -1 & 2 \\ 3 & -2 & 0 \end{pmatrix}$, and $C = \begin{pmatrix} 1 & 6 \\ -2 & 4 \\ 0 & 5 \end{pmatrix}$.

19. As in Example 3, suppose that a group of people have contracted a contagious disease. These persons have contacts with a second group who in turn have contacts with a third group. Let $A = \begin{pmatrix} 1 & 0 & 1 & 0 \\ 0 & 1 & 1 & 0 \\ 1 & 0 & 0 & 1 \end{pmatrix}$ represent the contacts between the contagious group and the members of group 2, and let

$$B = \begin{pmatrix} 1 & 0 & 1 & 0 & 0 \\ 0 & 0 & 0 & 1 & 0 \\ 1 & 1 & 0 & 0 & 0 \\ 0 & 0 & 1 & 0 & 1 \end{pmatrix}$$

represent the contacts between groups 2 and 3. (a) How many people are in each group? (b) Find the matrix of indirect contacts between groups 1 and 3.

20. Answer the questions of Problem 19 for $A = \begin{pmatrix} 1 & 0 & 1 & 1 & 0 \\ 0 & 1 & 0 & 1 & 1 \end{pmatrix}$ and

$$B = \begin{pmatrix} 1 & 0 & 0 & 0 & 0 & 0 & 1 \\ 0 & 1 & 0 & 1 & 0 & 0 & 0 \\ 1 & 1 & 0 & 0 & 1 & 1 & 1 \\ 0 & 0 & 0 & 1 & 1 & 0 & 1 \\ 0 & 1 & 0 & 0 & 0 & 0 & 0 \end{pmatrix}$$

21. A company pays its executives a salary and gives them shares of its stock as an annual bonus. Last year, the president of the company received $80,000 and 50 shares of stock, each of the three vice-presidents were paid $45,000 and 20 shares of stock, and the treasurer was paid $40,000 and 10 shares of stock.
    (a) Express the payments to the executives in money and stock by means of a 2 × 3 matrix.
    (b) Express the number of executives of each rank by means of a column vector.
    (c) Use matrix multiplication to calculate the total amount of money and the total number of shares of stock the company paid these executives last year.

22. Sales, unit gross profits, and unit taxes for sales of a large corporation are given in the table below.

Month	Product Sales of Item			Item	Unit Profit (in hundreds of dollars)	Unit Taxes (in hundreds of dollars)
	I	II	III			
January	4	2	20	I	3.5	1.5
February	6	1	9	II	2.75	2
March	5	3	12	III	1.5	0.6
April	8	2.5	20			

Find a matrix that shows total profits and taxes in each of the 4 months.

23. Let A be a square matrix. Then A^2 is defined simply as AA. Calculate $\begin{pmatrix} 2 & -1 \\ 4 & 6 \end{pmatrix}^2$.

24. Calculate A^2, where $A = \begin{pmatrix} 1 & -2 & 4 \\ 2 & 0 & 3 \\ 1 & 1 & 5 \end{pmatrix}$.

25. Calculate A^3, where $A = \begin{pmatrix} -1 & 2 \\ 3 & 4 \end{pmatrix}$.

26. Calculate A^2, A^3, A^4, and A^5, where

$$A = \begin{pmatrix} 0 & 1 & 0 & 0 \\ 0 & 0 & 1 & 0 \\ 0 & 0 & 0 & 1 \\ 0 & 0 & 0 & 0 \end{pmatrix}$$

27. Calculate A^2, A^3, A^4, and A^5, where

$$A = \begin{pmatrix} 0 & 1 & 0 & 0 & 0 \\ 0 & 0 & 1 & 0 & 0 \\ 0 & 0 & 0 & 1 & 0 \\ 0 & 0 & 0 & 0 & 1 \\ 0 & 0 & 0 & 0 & 0 \end{pmatrix}$$

28. An $n \times n$ matrix A has the property that AB is the zero matrix for any $n \times n$ matrix B. Prove that A is the zero matrix.

29. A **probability matrix** is a square matrix having two properties: (i) every component is nonnegative (≥ 0) and (ii) the sum of the elements in each row is 1. The following are probability matrices:

$$P = \begin{pmatrix} \frac{1}{3} & \frac{1}{3} & \frac{1}{3} \\ \frac{1}{4} & \frac{1}{2} & \frac{1}{4} \\ 0 & 0 & 1 \end{pmatrix} \quad \text{and} \quad Q = \begin{pmatrix} \frac{1}{6} & \frac{1}{6} & \frac{2}{3} \\ 0 & 1 & 0 \\ \frac{1}{5} & \frac{1}{5} & \frac{3}{5} \end{pmatrix}$$

Show that PQ is a probability matrix.

* 30. Let P be a probability matrix. Show that P^2 is a probability matrix.

**31. Let P and Q be probability matrices of the same size. Prove that PQ is a probability matrix.

32. Prove formula (4) by using the associative law (Equation 3).

*33. A round robin tennis tournament can be organized in the following way. Each of the n players plays all the others, and the results are recorded in an $n \times n$ matrix R as follows:

$$R_{ij} = \begin{cases} 1 & \text{if the } i\text{th player beats the } j\text{th player} \\ 0 & \text{if the } i\text{th player loses to the } j\text{th player} \\ 0 & \text{if } i = j \end{cases}$$

The ith player is then assigned the score

$$S_i = \sum_{j=1}^{n} R_{ij} + \frac{1}{2} \sum_{j=1}^{n} (R^2)_{ij} \dagger$$

a. In a tournament between four players

$$R = \begin{pmatrix} 0 & 1 & 0 & 0 \\ 0 & 0 & 1 & 1 \\ 1 & 0 & 0 & 0 \\ 1 & 0 & 1 & 0 \end{pmatrix}.$$

Rank the players according to their scores.

b. Interpret the meaning of the score.

34. Let O be the $m \times n$ zero matrix and let A be an $n \times p$ matrix. Show that $OA = O_1$, where O_1 is the $m \times p$ zero matrix.

35. Verify the distributive law (Equation 5) for the matrices

$$A = \begin{pmatrix} 1 & 2 & 4 \\ 3 & -1 & 0 \end{pmatrix} \quad B = \begin{pmatrix} 2 & 7 \\ -1 & 4 \\ 0 & 0 \end{pmatrix} \quad C = \begin{pmatrix} -1 & 2 \\ 3 & 7 \\ 4 & 1 \end{pmatrix}$$

Since we shall be using the $\sum$ notation in several parts of this book, you should become familiar with it. The following problems are designed to help you with the symbol. In Problems 36–43 evaluate the given sums.

36. $\sum_{k=1}^{4} 2k$ **37.** $\sum_{i=1}^{3} i^3$ **38.** $\sum_{k=0}^{6} 1$ **39.** $\sum_{k=1}^{8} 3^k$

40. $\sum_{i=2}^{5} \frac{1}{1+i}$ **41.** $\sum_{j=5}^{7} \frac{2j+3}{j-2}$ **42.** $\sum_{i=1}^{3} \sum_{j=1}^{4} ij$ **43.** $\sum_{k=1}^{3} \sum_{j=2}^{4} k^2 j^3$

In Problems 44–56 write each sum using the $\sum$ notation.

44. $1 + 2 + 4 + 8 + 16$ **45.** $1 - 3 + 9 - 27 + 81 - 243$

46. $\frac{2}{3} + \frac{3}{4} + \frac{4}{5} + \frac{5}{6} + \frac{6}{7} + \frac{7}{8} + \cdots + \frac{n}{n+1}$

47. $1 + 2^{1/2} + 3^{1/3} + 4^{1/4} + 5^{1/5} + \cdots + n^{1/n}$

48. $1 + x^3 + x^6 + x^9 + x^{12} + x^{15} + x^{18} + x^{21}$

49. $-1 + \frac{1}{a} - \frac{1}{a^2} + \frac{1}{a^3} - \frac{1}{a^4} + \frac{1}{a^5} - \frac{1}{a^6} + \frac{1}{a^7} - \frac{1}{a^8} + \frac{1}{a^9}$

---

†$(R^2)_{ij}$ is the ijth component of the matrix R^2.

50. $1 \cdot 3 + 3 \cdot 5 + 5 \cdot 7 + 7 \cdot 9 + 9 \cdot 11 + 11 \cdot 13 + 13 \cdot 15 + 15 \cdot 17$

51. $2^2 \cdot 4 + 3^2 \cdot 6 + 4^2 \cdot 8 + 5^2 \cdot 10 + 6^2 \cdot 12 + 7^2 \cdot 14$

52. $a_{11} + a_{12} + a_{13} + a_{21} + a_{22} + a_{23}$

53. $a_{11} + a_{12} + a_{21} + a_{22} + a_{31} + a_{32}$

54. $a_{21} + a_{22} + a_{23} + a_{24} + a_{31} + a_{32} + a_{33} + a_{34} + a_{41} + a_{42} + a_{43} + a_{44}$

55. $a_{31}b_{12} + a_{32}b_{22} + a_{33}b_{32} + a_{34}b_{42} + a_{35}b_{52}$

56. $a_{21}b_{11}c_{15} + a_{21}b_{12}c_{25} + a_{21}b_{13}c_{35} + a_{21}b_{14}c_{45}$
$+ a_{22}b_{21}c_{15} + a_{22}b_{22}c_{25} + a_{22}b_{23}c_{35} + a_{22}b_{24}c_{45}$
$+ a_{23}b_{31}c_{15} + a_{23}b_{32}c_{25} + a_{23}b_{33}c_{35} + a_{23}b_{34}c_{45}$

2.5 Matrices and Linear Systems of Equations

In Section 1.3, page 15, we discussed the following systems of m equations in n unknowns:

$$\begin{aligned} a_{11}x_1 + a_{12}x_2 + \cdots + a_{1n}x_n &= b_1 \\ a_{21}x_1 + a_{22}x_2 + \cdots + a_{2n}x_n &= b_2 \\ \vdots \qquad \vdots \qquad \qquad \vdots & \quad \vdots \\ a_{m1}x_1 + a_{m2}x_2 + \cdots + a_{mn}x_n &= b_m \end{aligned} \qquad (1)$$

We define the matrix

$$A = \begin{pmatrix} a_{11} & a_{12} & \cdots & a_{1n} \\ a_{21} & a_{22} & \cdots & a_{2n} \\ \vdots & \vdots & & \vdots \\ a_{m1} & a_{m2} & \cdots & a_{mn} \end{pmatrix},$$

the vector $\mathbf{x} = \begin{pmatrix} x_1 \\ x_2 \\ \vdots \\ x_n \end{pmatrix}$, and the vector $\mathbf{b} = \begin{pmatrix} b_1 \\ b_2 \\ \vdots \\ b_m \end{pmatrix}$. Since A is an $m \times n$ matrix and $\mathbf{x}$ is an $n \times 1$ matrix, the matrix product $A\mathbf{x}$ is defined as an $m \times 1$ matrix. It is not difficult to see that system (1) can be written as

$$A\mathbf{x} = \mathbf{b} \qquad (2)$$

EXAMPLE 1   Consider the system

$$\begin{aligned} 2x_1 + 4x_2 + 6x_3 &= 18 \\ 4x_1 + 5x_2 + 6x_3 &= 24 \\ 3x_1 + x_2 - 2x_3 &= 4 \end{aligned} \qquad (3)$$

(See Example 1.3.1 on page 6.) This can be written in the form $A\mathbf{x} = \mathbf{b}$ with $A = \begin{pmatrix} 2 & 4 & 6 \\ 4 & 5 & 6 \\ 3 & 1 & -2 \end{pmatrix}$, $\mathbf{x} = \begin{pmatrix} x_1 \\ x_2 \\ x_3 \end{pmatrix}$, and $\mathbf{b} = \begin{pmatrix} 18 \\ 24 \\ 4 \end{pmatrix}$.

It is obviously easier to write out system (1) in the form $A\mathbf{x} = \mathbf{b}$. There are many other advantages, too. In Section 2.7 we shall see how a square system can be solved almost at once if we know a matrix called the *inverse* of A. Even without that, as we saw in Chapter 1, computations are much easier to write down by using an augmented matrix. Let us repeat the computations of Example 1.3.1 starting with the augmented matrix:

$$\begin{pmatrix} 2 & 4 & 6 & | & 18 \\ 4 & 5 & 6 & | & 24 \\ 3 & 1 & -2 & | & 4 \end{pmatrix} \xrightarrow{M_1(\frac{1}{2})} \begin{pmatrix} 1 & 2 & 3 & | & 9 \\ 4 & 5 & 6 & | & 24 \\ 3 & 1 & -2 & | & 4 \end{pmatrix} \xrightarrow[A_{1,3}(-3)]{A_{1,2}(-4)} \begin{pmatrix} 1 & 2 & 3 & | & 9 \\ 0 & -3 & -6 & | & -12 \\ 0 & -5 & -11 & | & -23 \end{pmatrix}$$

$$\xrightarrow{M_2(-\frac{1}{3})} \begin{pmatrix} 1 & 2 & 3 & | & 9 \\ 0 & 1 & 2 & | & 4 \\ 0 & -5 & -11 & | & -23 \end{pmatrix} \xrightarrow[A_{2,3}(5)]{A_{2,1}(-2)} \begin{pmatrix} 1 & 0 & -1 & | & 1 \\ 0 & 1 & 2 & | & 4 \\ 0 & 0 & -1 & | & -3 \end{pmatrix}$$

$$\xrightarrow{M_3(-1)} \begin{pmatrix} 1 & 0 & -1 & | & 1 \\ 0 & 1 & 2 & | & 4 \\ 0 & 0 & 1 & | & 3 \end{pmatrix} \xrightarrow[A_{3,2}(-2)]{A_{3,1}(1)} \begin{pmatrix} 1 & 0 & 0 & | & 4 \\ 0 & 1 & 0 & | & -2 \\ 0 & 0 & 1 & | & 3 \end{pmatrix}$$

The last augmented matrix tells us that $x_1 = 4$, $x_2 = -2$, and $x_3 = 3$, as we already knew.

In this last example it is important to note that the last system of equations can be written as

$$I\mathbf{x} = \mathbf{s} \tag{4}$$

where $I = \begin{pmatrix} 1 & 0 & 0 \\ 0 & 1 & 0 \\ 0 & 0 & 1 \end{pmatrix}$ and $\mathbf{s}$ is the solution vector $\begin{pmatrix} 4 \\ -2 \\ 3 \end{pmatrix}$. We shall be making use of this fact in Section 2.7.

PROBLEMS 2.5 In Problems 1–6 write the given system in the form $A\mathbf{x} = \mathbf{b}$.

1. $2x_1 - x_2 = 3$
   $4x_1 + 5x_2 = 7$

2. $x_1 - x_2 + 3x_3 = 11$
   $4x_1 + x_2 - x_3 = -4$
   $2x_1 - x_2 + 3x_3 = 10$

3. $3x_1 + 6x_2 - 7x_3 = 0$
   $2x_1 - x_2 + 3x_3 = 1$

4. $4x_1 - x_2 + x_3 - x_4 = -7$
   $3x_1 + x_2 - 5x_3 + 6x_4 = 8$
   $2x_1 - x_2 + x_3 = 9$

5. $\quad\quad x_2 - x_3 = 7$
   $x_1 \quad\quad + x_3 = 2$
   $3x_1 + 2x_2 \quad\quad = -5$

6. $2x_1 + 3x_2 - x_3 = 0$
   $-4x_1 + 2x_2 + x_3 = 0$
   $7x_1 + 3x_2 - 9x_3 = 0$

In Problems 7–15 write out the system of equations represented by the given augmented matrix.

7. $\begin{pmatrix} 1 & 1 & -1 & | & 7 \\ 4 & -1 & 5 & | & 4 \\ 6 & 1 & 3 & | & 20 \end{pmatrix}$

8. $\begin{pmatrix} 0 & 1 & | & 2 \\ 1 & 0 & | & 3 \end{pmatrix}$

9. $\begin{pmatrix} 2 & 0 & 1 & | & 2 \\ -3 & 4 & 0 & | & 3 \\ 0 & 5 & 6 & | & 5 \end{pmatrix}$

10. $\begin{pmatrix} 2 & 3 & 1 & | & 2 \\ 0 & 4 & 1 & | & 3 \\ 0 & 0 & 0 & | & 0 \end{pmatrix}$

11. $\begin{pmatrix} 1 & 0 & 0 & 0 & | & 2 \\ 0 & 1 & 0 & 0 & | & 3 \\ 0 & 0 & 1 & 0 & | & -5 \\ 0 & 0 & 0 & 1 & | & 6 \end{pmatrix}$

12. $\begin{pmatrix} 2 & 3 & 1 & | & 0 \\ 4 & -1 & 5 & | & 0 \\ 3 & 6 & -7 & | & 0 \end{pmatrix}$

13. $\begin{pmatrix} 6 & 2 & 1 & | & 2 \\ -2 & 3 & 1 & | & 4 \\ 0 & 0 & 0 & | & 2 \end{pmatrix}$

14. $\begin{pmatrix} 3 & 1 & 5 & | & 6 \\ 2 & 3 & 2 & | & 4 \end{pmatrix}$

15. $\begin{pmatrix} 7 & 2 & | & 1 \\ 3 & 1 & | & 2 \\ 6 & 9 & | & 3 \end{pmatrix}$

16. Solve the system represented by the augmented matrix of Problem 9.

17. Solve the system represented by $\begin{pmatrix} 1 & 2 & -4 & | & 4 \\ -2 & -4 & 8 & | & -8 \end{pmatrix}$.

18. Solve the system represented by $\begin{pmatrix} 1 & 2 & -4 & | & 4 \\ -2 & -4 & 8 & | & -9 \end{pmatrix}$.

19. Solve the homogeneous system represented by $\begin{pmatrix} 1 & -2 & 3 & | & 0 \\ 4 & 1 & -1 & | & 0 \\ 2 & -1 & 3 & | & 0 \end{pmatrix}$.

20. Solve the homogeneous system represented by $\begin{pmatrix} 1 & 1 & -1 & | & 0 \\ 4 & -1 & 5 & | & 0 \\ 6 & 1 & 3 & | & 0 \end{pmatrix}$.

21. Solve the system represented by the augmented matrix

$$\begin{pmatrix} 1 & 3 & -2 & 1 & | & 3 \\ 2 & -6 & 4 & -1 & | & 2 \\ 4 & 12 & -8 & 2 & | & 4 \\ -3 & 0 & 6 & -2 & | & -8 \end{pmatrix}$$

22. Solve the homogeneous system represented by the augmented matrix

$$\begin{pmatrix} 1 & 2 & -3 & 5 & 4 & | & 0 \\ -2 & 4 & 7 & -3 & 5 & | & 0 \\ -4 & 0 & 13 & -13 & -3 & | & 0 \end{pmatrix}$$

23. Find a matrix A and vectors $\mathbf{x}$ and $\mathbf{b}$ such that the system represented by the following augmented matrix can be written in the form $A\mathbf{x} = \mathbf{b}$ and solve the system.

$$\begin{pmatrix} 2 & 0 & 0 & | & 3 \\ 0 & 4 & 0 & | & 5 \\ 0 & 0 & -5 & | & 2 \end{pmatrix}$$

2.6 Linear Independence and Homogeneous Systems

In the study of linear algebra, one of the central ideas is that of the linear dependence or independence of vectors. In this section we shall define what we mean by linear independence and show how it is related to the theory of homogeneous systems of equations. We return to linear independence in Chapters 3 and 4 and, in Chapter 5, we shall see how this concept is central to the theory of vector spaces.

Is there a special relationship between the vectors $\mathbf{v}_1 = \begin{pmatrix} 1 \\ 2 \end{pmatrix}$ and $\mathbf{v}_2 = \begin{pmatrix} 2 \\ 4 \end{pmatrix}$? Of course, we see that $\mathbf{v}_2 = 2\mathbf{v}_1$ or, writing this equation in another way,

$$2\mathbf{v}_1 - \mathbf{v}_2 = \mathbf{0} \tag{1}$$

What is special about the vectors $\mathbf{v}_1 = \begin{pmatrix} 1 \\ 2 \\ 3 \end{pmatrix}$, $\mathbf{v}_2 = \begin{pmatrix} -4 \\ 1 \\ 5 \end{pmatrix}$, and $\mathbf{v}_3 = \begin{pmatrix} -5 \\ 8 \\ 19 \end{pmatrix}$? This question is more difficult to answer at first glance. It is easy to verify, however, that $\mathbf{v}_3 = 3\mathbf{v}_1 + 2\mathbf{v}_2$, or, rewriting,

$$3\mathbf{v}_1 + 2\mathbf{v}_2 - \mathbf{v}_3 = \mathbf{0} \tag{2}$$

It appears that the two vectors in Equation (1) and the three vectors in (2) are more closely related than an arbitrary pair of 2-vectors or an arbitrary triple of 3-vectors. In each case we say that the vectors are *linearly dependent*.[†] In general, we have the following important definition.

DEFINITION 1   **LINEARLY DEPENDENT VECTORS**   The set of vectors $\mathbf{v}_1, \mathbf{v}_2, \ldots, \mathbf{v}_n$ is **linearly dependent** if there exist scalars $c_1, c_2, \ldots, c_n$ *not all zero* such that

$$c_1\mathbf{v}_1 + c_2\mathbf{v}_2 + \cdots + c_n\mathbf{v}_n = \mathbf{0} \tag{3}$$

With this definition we see that the vectors in Equation (1) $[c_1 = 2, c_2 = -1]$ and Equation (2) $[c_1 = 3, c_2 = 2, c_3 = -1]$ are linearly dependent.

DEFINITION 2   **LINEARLY INDEPENDENT VECTORS**   The set of vectors $\mathbf{v}_1, \mathbf{v}_2, \ldots, \mathbf{v}_n$ is **linearly independent** if it is not linearly dependent.

Putting this another way, $\mathbf{v}_1, \mathbf{v}_2, \ldots, \mathbf{v}_n$ are linearly independent if the equation $c_1\mathbf{v}_1 + c_2\mathbf{v}_2 + \cdots + c_n\mathbf{v}_n = \mathbf{0}$ holds only for $c_1 = c_2 = \cdots = c_n = 0$.

How do we determine whether a set of vectors is linearly dependent or independent? The case for two vectors is easy.

[†] In Chapter 4 (Sections 4.1 and 4.3) we shall see that two linearly dependent vectors in the xy-plane are collinear and that three linearly dependent vectors in space are coplanar.

2/VECTORS AND MATRICES

THEOREM 1   Two vectors are linearly dependent if and only if one is a scalar multiple of the other.

Proof   First suppose that $\mathbf{v}_2 = c\mathbf{v}_1$ for some scalar $c \neq 0$. Then $c\mathbf{v}_1 - \mathbf{v}_2 = \mathbf{0}$ and $\mathbf{v}_1$ and $\mathbf{v}_2$ are linearly dependent. On the other hand, suppose that $\mathbf{v}_1$ and $\mathbf{v}_2$ are dependent. Then there are constants c_1 and c_2, not both zero, such that $c_1\mathbf{v}_1 + c_2\mathbf{v}_2 = \mathbf{0}$. If $c_1 \neq 0$, then, dividing by c_1, we obtain $\mathbf{v}_1 + (c_2/c_1)\mathbf{v}_2 = \mathbf{0}$ or

$$\mathbf{v}_1 = \left(-\frac{c_2}{c_1}\right)\mathbf{v}_2$$

That is, $\mathbf{v}_1$ is a scalar multiple of $\mathbf{v}_2$. If $c_1 = 0$ then $c_2 \neq 0$ and hence $\mathbf{v}_2 = \mathbf{0} = 0\mathbf{v}_1$. ∎

EXAMPLE 1   The vectors $\mathbf{v}_1 = \begin{pmatrix} 2 \\ -1 \\ 0 \\ 3 \end{pmatrix}$ and $\mathbf{v}_2 = \begin{pmatrix} -6 \\ 3 \\ 0 \\ -9 \end{pmatrix}$ are linearly dependent since $\mathbf{v}_2 = -3\mathbf{v}_1$.

EXAMPLE 2   The vectors $\begin{pmatrix} 1 \\ 2 \\ 4 \end{pmatrix}$ and $\begin{pmatrix} 2 \\ 5 \\ -3 \end{pmatrix}$ are linearly independent; if they were not, we would have $\begin{pmatrix} 2 \\ 5 \\ -3 \end{pmatrix} = c\begin{pmatrix} 1 \\ 2 \\ 4 \end{pmatrix} = \begin{pmatrix} c \\ 2c \\ 4c \end{pmatrix}$. Then $2 = c$, $5 = 2c$, and $-3 = 4c$, which is clearly impossible for any number c.

There are several techniques for determining whether a set of vectors is linearly independent. Let us examine one of these techniques here. Another is given in Chapter 3 (see Example 3.4.6 on page 111).

EXAMPLE 3   Determine whether the vectors $\begin{pmatrix} 1 \\ -2 \\ 3 \end{pmatrix}$, $\begin{pmatrix} 2 \\ -2 \\ 0 \end{pmatrix}$, and $\begin{pmatrix} 0 \\ 1 \\ 7 \end{pmatrix}$ are linearly dependent or independent.

Solution   Suppose that $c_1 \begin{pmatrix} 1 \\ -2 \\ 3 \end{pmatrix} + c_2 \begin{pmatrix} 2 \\ -2 \\ 0 \end{pmatrix} + c_3 \begin{pmatrix} 0 \\ 1 \\ 7 \end{pmatrix} = \mathbf{0} = \begin{pmatrix} 0 \\ 0 \\ 0 \end{pmatrix}$. Then, multiplying through and adding, we have $\begin{pmatrix} c_1 + 2c_2 \\ -2c_1 - 2c_2 + c_3 \\ 3c_1 + 7c_3 \end{pmatrix} = \begin{pmatrix} 0 \\ 0 \\ 0 \end{pmatrix}$. This yields a

2.6 / LINEAR INDEPENDENCE AND HOMOGENEOUS SYSTEMS

homogeneous system of three equations in the three unknowns c_1, c_2, and c_3:

$$\begin{aligned} c_1 + 2c_2 &= 0 \\ -2c_1 - 2c_2 + c_3 &= 0 \\ 3c_1 \quad\quad + 7c_3 &= 0 \end{aligned} \tag{4}$$

Thus the vectors will be linearly dependent if and only if system (4) has nontrivial solutions. We write system (4) using an augmented matrix and then row-reduce:

$$\begin{pmatrix} 1 & 2 & 0 & | & 0 \\ -2 & -2 & 1 & | & 0 \\ 3 & 0 & 7 & | & 0 \end{pmatrix} \xrightarrow{\substack{A_{1,2}(2) \\ A_{1,3}(-3)}} \begin{pmatrix} 1 & 2 & 0 & | & 0 \\ 0 & 2 & 1 & | & 0 \\ 0 & -6 & 7 & | & 0 \end{pmatrix}$$

$$\xrightarrow{M_2(\frac{1}{2})} \begin{pmatrix} 1 & 2 & 0 & | & 0 \\ 0 & 1 & \frac{1}{2} & | & 0 \\ 0 & -6 & 7 & | & 0 \end{pmatrix} \xrightarrow{\substack{A_{2,1}(-2) \\ A_{2,3}(6)}} \begin{pmatrix} 1 & 0 & -1 & | & 0 \\ 0 & 1 & \frac{1}{2} & | & 0 \\ 0 & 0 & 10 & | & 0 \end{pmatrix}$$

$$\xrightarrow{M_3(\frac{1}{10})} \begin{pmatrix} 1 & 0 & -1 & | & 0 \\ 0 & 1 & \frac{1}{2} & | & 0 \\ 0 & 0 & 1 & | & 0 \end{pmatrix} \xrightarrow{\substack{A_{3,1}(1) \\ A_{3,2}(-\frac{1}{2})}} \begin{pmatrix} 1 & 0 & 0 & | & 0 \\ 0 & 1 & 0 & | & 0 \\ 0 & 0 & 1 & | & 0 \end{pmatrix}$$

The last system of equations reads $c_1 = 0$, $c_2 = 0$, $c_3 = 0$. Hence (4) has no nontrivial solutions and the given vectors are linearly independent.

EXAMPLE 4 Determine whether the vectors $\begin{pmatrix} 1 \\ -3 \\ 0 \end{pmatrix}$, $\begin{pmatrix} 3 \\ 0 \\ 4 \end{pmatrix}$, and $\begin{pmatrix} 11 \\ -6 \\ 12 \end{pmatrix}$ are linearly dependent or independent.

Solution The equation $c_1 \begin{pmatrix} 1 \\ -3 \\ 0 \end{pmatrix} + c_2 \begin{pmatrix} 3 \\ 0 \\ 4 \end{pmatrix} + c_3 \begin{pmatrix} 11 \\ -6 \\ 12 \end{pmatrix} = \begin{pmatrix} 0 \\ 0 \\ 0 \end{pmatrix}$ leads to the homogeneous system

$$\begin{aligned} c_1 + 3c_2 + 11c_3 &= 0 \\ -3c_1 \quad\quad - 6c_3 &= 0 \\ 4c_2 + 12c_3 &= 0 \end{aligned} \tag{5}$$

Writing system (5) in augmented matrix form and row reducing we obtain, successively,

$$\begin{pmatrix} 1 & 3 & 11 & | & 0 \\ -3 & 0 & -6 & | & 0 \\ 0 & 4 & 12 & | & 0 \end{pmatrix} \xrightarrow{A_{1,2}(3)} \begin{pmatrix} 1 & 3 & 11 & | & 0 \\ 0 & 9 & 27 & | & 0 \\ 0 & 4 & 12 & | & 0 \end{pmatrix}$$

$$\xrightarrow{M_2(\frac{1}{9})} \begin{pmatrix} 1 & 3 & 11 & | & 0 \\ 0 & 1 & 3 & | & 0 \\ 0 & 4 & 12 & | & 0 \end{pmatrix} \xrightarrow{\substack{A_{2,1}(-3) \\ A_{2,3}(-4)}} \begin{pmatrix} 1 & 0 & 2 & | & 0 \\ 0 & 1 & 3 & | & 0 \\ 0 & 0 & 0 & | & 0 \end{pmatrix}$$

We can stop here since the theory of Section 1.4 shows us that system (5) has an infinite number of solutions. For example, the last augmented matrix reads

$$c_1 + 2c_3 = 0$$
$$c_2 + 3c_3 = 0$$

If we choose $c_3 = 1$, we have $c_2 = -3$ and $c_1 = -2$ so that, as is easily verified,

$$-2\begin{pmatrix} 1 \\ -3 \\ 0 \end{pmatrix} - 3\begin{pmatrix} 3 \\ 0 \\ 4 \end{pmatrix} + \begin{pmatrix} 11 \\ -6 \\ 12 \end{pmatrix} = \begin{pmatrix} 0 \\ 0 \\ 0 \end{pmatrix}$$

and the vectors are linearly dependent.

The theory of homogeneous systems can tell us something about the linear dependence or independence of vectors.

THEOREM 2 A set of n m-vectors is always linearly dependent if $n > m$.

Proof Let $v_1, v_2, \ldots, v_n$ be n m-vectors and let us try to find constants $c_1, c_2, \ldots, c_n$ not all zero such that

$$c_1 v_1 + c_2 v_2 + \cdots + c_n v_n = 0 \tag{6}$$

Let $v_1 = \begin{pmatrix} a_{11} \\ a_{21} \\ \vdots \\ a_{m1} \end{pmatrix}$, $v_2 = \begin{pmatrix} a_{12} \\ a_{22} \\ \vdots \\ a_{m2} \end{pmatrix}, \ldots, v_n = \begin{pmatrix} a_{1n} \\ a_{2n} \\ \vdots \\ a_{mn} \end{pmatrix}$. Then Equation (6) becomes

$$\begin{aligned} a_{11}c_1 + a_{12}c_2 + \cdots + a_{1n}c_n &= 0 \\ a_{21}c_1 + a_{22}c_2 + \cdots + a_{2n}c_n &= 0 \\ &\vdots \\ a_{m1}c_1 + a_{m2}c_2 + \cdots + a_{mn}c_n &= 0 \end{aligned} \tag{7}$$

But system (7) is system (1.4.1) on page 22 and, according to Theorem 1.4.1, this system has an infinite number of solutions if $n > m$. Thus there are scalars $c_1, c_2, \ldots, c_n$ not all zero that satisfy (7) and the vectors $v_1, v_2, \ldots, v_n$ are therefore linearly dependent. ■

EXAMPLE 5 The vectors $\begin{pmatrix} 2 \\ -3 \\ 4 \end{pmatrix}, \begin{pmatrix} 4 \\ 7 \\ -6 \end{pmatrix}, \begin{pmatrix} 18 \\ -11 \\ 4 \end{pmatrix}$, and $\begin{pmatrix} 2 \\ -7 \\ 3 \end{pmatrix}$ are linearly dependent since they comprise a set of four 3-vectors.

There is a very important (and obvious) corollary to Theorem 2.

COROLLARY A set of linearly independent n-vectors contains at most n vectors.

Note. We can rephrase the corollary as follows: If we have n linearly independent n-vectors, then we cannot add any more vectors without

2.6 / LINEAR INDEPENDENCE AND HOMOGENEOUS SYSTEMS

making the set linearly dependent. This fact will be very important in Chapter 5.

From system (7) we can make another important observation whose proof is left as an exercise (see Problem 17).

THEOREM 3 Let

$$A = \begin{pmatrix} a_{11} & a_{12} & \cdots & a_{1n} \\ a_{21} & a_{22} & \cdots & a_{2n} \\ \vdots & \vdots & & \vdots \\ a_{m1} & a_{m2} & \cdots & a_{mn} \end{pmatrix}$$

Then the columns of A, considered as vectors, are linearly dependent if and only if system (7), which can be written $A\mathbf{c} = \mathbf{0}$, has an infinite number of solutions. Here $\mathbf{c} = \begin{pmatrix} c_1 \\ c_2 \\ \vdots \\ c_n \end{pmatrix}$.

EXAMPLE 6 Consider the homogeneous system

$$\begin{aligned} x_1 + 2x_2 - x_3 + 2x_4 &= 0 \\ 3x_1 + 7x_2 + x_3 + 4x_4 &= 0 \end{aligned} \tag{8}$$

We solve this by row reduction:

$$\begin{pmatrix} 1 & 2 & -1 & 2 & | & 0 \\ 3 & 7 & 1 & 4 & | & 0 \end{pmatrix} \xrightarrow{A_{1,2}(-3)} \begin{pmatrix} 1 & 2 & -1 & 2 & | & 0 \\ 0 & 1 & 4 & -2 & | & 0 \end{pmatrix}$$

$$\xrightarrow{A_{2,1}(-2)} \begin{pmatrix} 1 & 0 & -9 & 6 & | & 0 \\ 0 & 1 & 4 & -2 & | & 0 \end{pmatrix}$$

The last system is

$$\begin{aligned} x_1 \quad\;\; - 9x_3 + 6x_4 &= 0 \\ x_2 + 4x_3 - 2x_4 &= 0 \end{aligned}$$

We see that this system has an infinite number of solutions, which we write as a column vector:

$$\begin{pmatrix} x_1 \\ x_2 \\ x_3 \\ x_4 \end{pmatrix} = \begin{pmatrix} 9x_3 - 6x_4 \\ -4x_3 + 2x_4 \\ x_3 \\ x_4 \end{pmatrix} = x_3 \begin{pmatrix} 9 \\ -4 \\ 1 \\ 0 \end{pmatrix} + x_4 \begin{pmatrix} -6 \\ 2 \\ 0 \\ 1 \end{pmatrix} \tag{9}$$

Note that $\begin{pmatrix} 9 \\ -4 \\ 1 \\ 0 \end{pmatrix}$ and $\begin{pmatrix} -6 \\ 2 \\ 0 \\ 1 \end{pmatrix}$ are linearly independent solutions to (8) because neither one is a multiple of the other. (You should verify that they are solutions.) Since x_3 and x_4 are arbitrary real numbers, we see, from (9), that we can express all solutions to the system (8) in terms of two linearly independent solution vectors.

PROBLEMS 2.6

In Problems 1–12 determine whether the given set of vectors is linearly dependent or independent.

1. $\begin{pmatrix} 1 \\ 2 \end{pmatrix}; \begin{pmatrix} -1 \\ -3 \end{pmatrix}$

2. $\begin{pmatrix} 2 \\ -1 \\ 4 \end{pmatrix}; \begin{pmatrix} 4 \\ -2 \\ 7 \end{pmatrix}$

3. $\begin{pmatrix} 2 \\ -1 \\ 4 \end{pmatrix}; \begin{pmatrix} 4 \\ -2 \\ 8 \end{pmatrix}$

4. $\begin{pmatrix} -2 \\ 3 \end{pmatrix}; \begin{pmatrix} 4 \\ 7 \end{pmatrix}$

5. $\begin{pmatrix} -3 \\ 2 \end{pmatrix}; \begin{pmatrix} 1 \\ 10 \end{pmatrix}; \begin{pmatrix} 4 \\ -5 \end{pmatrix}$

6. $\begin{pmatrix} 1 \\ 0 \\ 1 \end{pmatrix}; \begin{pmatrix} 0 \\ 1 \\ 1 \end{pmatrix}; \begin{pmatrix} 1 \\ 1 \\ 0 \end{pmatrix}$

7. $\begin{pmatrix} 1 \\ 0 \\ 0 \end{pmatrix}; \begin{pmatrix} 0 \\ 1 \\ 0 \end{pmatrix}; \begin{pmatrix} 0 \\ 0 \\ 1 \end{pmatrix}$

8. $\begin{pmatrix} -3 \\ 4 \\ 2 \end{pmatrix}; \begin{pmatrix} 7 \\ -1 \\ 3 \end{pmatrix}; \begin{pmatrix} 1 \\ 2 \\ 8 \end{pmatrix}$

9. $\begin{pmatrix} -3 \\ 4 \\ 2 \end{pmatrix}; \begin{pmatrix} 7 \\ -1 \\ 3 \end{pmatrix}; \begin{pmatrix} 1 \\ 1 \\ 8 \end{pmatrix}$

10. $\begin{pmatrix} 1 \\ -2 \\ 1 \\ 1 \end{pmatrix}; \begin{pmatrix} 3 \\ 0 \\ 2 \\ -2 \end{pmatrix}; \begin{pmatrix} 0 \\ 4 \\ -1 \\ -1 \end{pmatrix}; \begin{pmatrix} 5 \\ 0 \\ 3 \\ -1 \end{pmatrix}$

11. $\begin{pmatrix} 1 \\ -2 \\ 1 \\ 1 \end{pmatrix}; \begin{pmatrix} 3 \\ 0 \\ 2 \\ -2 \end{pmatrix}; \begin{pmatrix} 0 \\ 4 \\ -1 \\ 1 \end{pmatrix}; \begin{pmatrix} 5 \\ 0 \\ 3 \\ -1 \end{pmatrix}$

12. $\begin{pmatrix} 1 \\ -1 \\ 2 \end{pmatrix}; \begin{pmatrix} 4 \\ 0 \\ 0 \end{pmatrix}; \begin{pmatrix} -2 \\ 3 \\ 5 \end{pmatrix}; \begin{pmatrix} 7 \\ 1 \\ 2 \end{pmatrix}$

13. Determine a condition on the numbers a, b, c, and d such that the vectors $\begin{pmatrix} a \\ b \end{pmatrix}$ and $\begin{pmatrix} c \\ d \end{pmatrix}$ are linearly dependent.

*14. Find a condition on the numbers a_{ij} such that the vectors $\begin{pmatrix} a_{11} \\ a_{21} \\ a_{31} \end{pmatrix}$, $\begin{pmatrix} a_{12} \\ a_{22} \\ a_{32} \end{pmatrix}$, and $\begin{pmatrix} a_{13} \\ a_{23} \\ a_{33} \end{pmatrix}$ are linearly dependent.

15. For what value(s) of α will the vectors $\begin{pmatrix} 1 \\ 2 \\ 3 \end{pmatrix}, \begin{pmatrix} 2 \\ -1 \\ 4 \end{pmatrix}, \begin{pmatrix} 3 \\ \alpha \\ 4 \end{pmatrix}$ be linearly dependent?

16. For what value(s) of α are the vectors $\begin{pmatrix} 2 \\ -3 \\ 1 \end{pmatrix}, \begin{pmatrix} -4 \\ 6 \\ -2 \end{pmatrix}, \begin{pmatrix} \alpha \\ 1 \\ 2 \end{pmatrix}$ linearly dependent? [*Hint:* Look carefully.]

17. Prove Theorem 3. [*Hint:* Look closely at system (7).]

18. Prove that if the vectors $v_1, v_2, \ldots, v_n$ are linearly dependent m-vectors and if v_{n+1} is any other m-vector, then the set $v_1, v_2, \ldots, v_n, v_{n+1}$ is linearly dependent.
19. Show that if $v_1, v_2, \ldots, v_n$ $(n \geq 2)$ are linearly independent, then so too are $v_1, v_2, \ldots, v_k$, where $k < n$.
20. Show that if the nonzero vectors v_1 and v_2 are orthogonal (see Problem 2.2.15 page 36), then the set $\{v_1, v_2\}$ is linearly independent.
*21. Suppose that v_1 is orthogonal to v_2 and v_3 and that v_2 is orthogonal to v_3. If v_1, v_2, and v_3 are nonzero, show that the set $\{v_1, v_2, v_3\}$ is linearly independent.
22. Let A be a square matrix whose columns are the vectors $v_1, v_2, \ldots, v_n$. Show that $v_1, v_2, \ldots, v_n$ are linearly independent if and only if the row echelon form of A does not contain a row of zeros.

In Problems 23–26 write the solutions to the given homogeneous systems in terms of one or more linearly independent vectors.

23. $x_1 + x_2 + x_3 = 0$
24. $x_1 - x_2 + 7x_3 - x_4 = 0$
    $2x_1 + 3x_2 - 8x_3 + x_4 = 0$
25. $x_1 + 2x_2 - x_3 = 0$
    $2x_1 + 5x_2 + 4x_3 = 0$
26. $x_1 + x_2 + x_3 - x_4 - x_5 = 0$
    $-2x_1 + 3x_2 + x_3 + 4x_4 - 6x_5 = 0$

2.7 The Inverse of a Square Matrix

In this section we define two kinds of matrices that are central to matrix theory. We begin with a simple example. Let $A = \begin{pmatrix} 2 & 5 \\ 1 & 3 \end{pmatrix}$ and $B = \begin{pmatrix} 3 & -5 \\ -1 & 2 \end{pmatrix}$. Then an easy computation shows that $AB = BA = I_2$, where $I_2 = \begin{pmatrix} 1 & 0 \\ 0 & 1 \end{pmatrix}$. The matrix I_2 is called the 2×2 *identity matrix*. The matrix B is called the *inverse* of A and is written A^{-1}.

DEFINITION 1   **IDENTITY MATRIX**   The $n \times n$ **identity matrix** is the $n \times n$ matrix with 1's down the **main diagonal**† and 0's everywhere else. That is,

$$I_n = (b_{ij}) \quad \text{where} \quad b_{ij} = \begin{cases} 1 & \text{if } i = j \\ 0 & \text{if } i \neq j \end{cases} \tag{1}$$

---
†The main diagonal of $A = (a_{ij})$ consists of the components a_{11}, a_{22}, a_{33}, and so on. Unless otherwise stated, we shall refer to the main diagonal simply as the **diagonal**.

EXAMPLE 1 $I_3 = \begin{pmatrix} 1 & 0 & 0 \\ 0 & 1 & 0 \\ 0 & 0 & 1 \end{pmatrix}$ and $I_5 = \begin{pmatrix} 1 & 0 & 0 & 0 & 0 \\ 0 & 1 & 0 & 0 & 0 \\ 0 & 0 & 1 & 0 & 0 \\ 0 & 0 & 0 & 1 & 0 \\ 0 & 0 & 0 & 0 & 1 \end{pmatrix}$

THEOREM 1 Let A be a square $n \times n$ matrix. Then

$$AI_n = I_n A = A$$

That is, I_n commutes with every $n \times n$ matrix and leaves it unchanged after multiplication on the left or right.

Note. I_n functions for $n \times n$ matrices the way the number 1 functions for real numbers (since $1 \cdot a = a \cdot 1 = a$ for every real number a).

Proof Let c_{ij} be the ijth element of AI_n. Then

$$c_{ij} = a_{i1}b_{1j} + a_{i2}b_{2j} + \cdots + a_{ij}b_{jj} + \cdots + a_{in}b_{nj}.$$

But, from (1), this sum is equal to a_{ij}. Thus $AI_n = A$. In a similar fashion we can show that $I_n A = A$, and this proves the theorem. ∎

Notation From now on we shall write the identity matrix simply as I, since if A is $n \times n$, the products IA and AI are defined only if I is also $n \times n$.

DEFINITION 2 **THE INVERSE OF A MATRIX** Let A and B be $n \times n$ matrices. Suppose that

$$AB = BA = I$$

Then B is called the **inverse** of A and is written as A^{-1}. We then have

$$AA^{-1} = A^{-1}A = I$$

If A has an inverse, then A is said to be **invertible**.

Remark 1. From this definition it immediately follows that $(A^{-1})^{-1} = A$ if A is invertible.

Remark 2. This definition does *not* state that every square matrix has an inverse. In fact there are many square matrices that have no inverse. (See, for instance, Example 2 below.)

In Definition 2 we defined *the* inverse of a matrix. This statement suggests that inverses are unique. This is indeed the case, as the following theorem shows.

THEOREM 2   If a square matrix A is invertible, then its inverse is unique.

Proof   Suppose B and C are two inverses for A. We can show that $B = C$. By definition, we have $AB = BA = I$ and $AC = CA = I$. Then $B(AC) = BI = B$ and $(BA)C = IC = C$. But $B(AC) = (BA)C$ by the associative law of matrix multiplication. Hence $B = C$ and the theorem is proved. ∎

Another important fact about inverses is given below.

THEOREM 3   Let A and B be invertible $n \times n$ matrices. Then AB is invertible and

$$(AB)^{-1} = B^{-1}A^{-1}$$

Proof   To prove this result, we refer to Definition 2. That is, $B^{-1}A^{-1} = (AB)^{-1}$ if and only if $B^{-1}A^{-1}(AB) = (AB)(B^{-1}A^{-1}) = I$. But this follows since

and
$$(B^{-1}A^{-1})(AB) \overset{\text{Equation (4) on page 45}}{=} B^{-1}(A^{-1}A)B = B^{-1}IB = B^{-1}B = I$$
$$(AB)(B^{-1}A^{-1}) = A(BB^{-1})A^{-1} = AIA^{-1} = AA^{-1} = I. \quad \blacksquare$$

Consider the system of n equations in n unknowns

$$A\mathbf{x} = \mathbf{b},$$

and suppose that A is invertible. Then

$$A^{-1}A\mathbf{x} = A^{-1}\mathbf{b} \quad \text{we multiplied on the left by } A^{-1}$$
$$I\mathbf{x} = A^{-1}\mathbf{b} \quad A^{-1}A = I$$
$$\mathbf{x} = A^{-1}\mathbf{b} \quad I\mathbf{x} = \mathbf{x}$$

That is,

If A is invertible, the system $A\mathbf{x} = \mathbf{b}$ has the unique solution $\mathbf{x} = A^{-1}\mathbf{b}.$ (2)

This is one of the reasons we study matrix inverses.

There are two basic questions that come to mind once we have defined the inverse of a matrix.

Question 1.   What matrices do have inverses?

Question 2.   If a matrix has an inverse, how can we compute it?

We answer both questions in this section. Rather than starting by giving you what seems to be a set of arbitrary rules, we look first at what happens in the 2×2 case.

EXAMPLE 2 Let $A = \begin{pmatrix} 2 & -3 \\ -4 & 5 \end{pmatrix}$. Compute A^{-1} if it exists.

Solution Suppose that A^{-1} exists. We write $A^{-1} = \begin{pmatrix} x & y \\ z & w \end{pmatrix}$ and use the fact that $AA^{-1} = I$. Then

$$AA^{-1} = \begin{pmatrix} 2 & -3 \\ -4 & 5 \end{pmatrix}\begin{pmatrix} x & y \\ z & w \end{pmatrix} = \begin{pmatrix} 2x - 3z & 2y - 3w \\ -4x + 5z & -4y + 5w \end{pmatrix} = \begin{pmatrix} 1 & 0 \\ 0 & 1 \end{pmatrix}$$

The last two matrices can be equal only if each of their corresponding components are equal. This means that

$$2x \quad - 3z \quad = 1 \tag{3}$$
$$2y \quad - 3w = 0 \tag{4}$$
$$-4x \quad + 5z \quad = 0 \tag{5}$$
$$-4y \quad + 5w = 1 \tag{6}$$

This is a system of four equations in four unknowns. Note that there are two equations involving x and z only (equations (3) and (5)) and two equations involving y and w only (equations (4) and (6)). We write these two systems in augmented matrix form:

$$\begin{pmatrix} 2 & -3 & | & 1 \\ -4 & 5 & | & 0 \end{pmatrix} \tag{7}$$

$$\begin{pmatrix} 2 & -3 & | & 0 \\ -4 & 5 & | & 1 \end{pmatrix}. \tag{8}$$

Now, we know from Section 1.3 that if system (7) (in the variables x and z) has a unique solution, then Gauss-Jordan elimination of (7) will result in

$$\begin{pmatrix} 1 & 0 & | & x \\ 0 & 1 & | & z \end{pmatrix}$$

where (x, z) is the unique pair of numbers that satisfies $2x - 3z = 1$ and $-4x + 5z = 0$. Similarly, row reduction of (8) will result in

$$\begin{pmatrix} 1 & 0 & | & y \\ 0 & 1 & | & w \end{pmatrix}$$

where (y, w) is the unique pair of numbers that satisfies $2y - 3w = 0$ and $-4y + 5w = 1$.

Since the coefficient matrices in (7) and (8) are the same, we can perform the row reductions on the two augmented matrices simultaneously, by considering the new augmented matrix

$$\begin{pmatrix} 2 & -3 & | & 1 & 0 \\ -4 & 5 & | & 0 & 1 \end{pmatrix}. \quad (9)$$

If A^{-1} is invertible, then the system defined by (3), (4), (5), and (6) has a unique solution and, by what we said above, Gauss-Jordan elimination will result in

$$\begin{pmatrix} 1 & 0 & | & x & y \\ 0 & 1 & | & z & w \end{pmatrix}.$$

We now carry out the computation, noting that the matrix on the left in (9) is A and the matrix on the right in (9) is I:

$$\begin{pmatrix} 2 & -3 & | & 1 & 0 \\ -4 & 5 & | & 0 & 1 \end{pmatrix} \xrightarrow{M_1(\frac{1}{2})} \begin{pmatrix} 1 & -\frac{3}{2} & | & \frac{1}{2} & 0 \\ -4 & 5 & | & 0 & 1 \end{pmatrix}$$

$$\xrightarrow{A_{1,2}(4)} \begin{pmatrix} 1 & -\frac{3}{2} & | & \frac{1}{2} & 0 \\ 0 & -1 & | & 2 & 1 \end{pmatrix}$$

$$\xrightarrow{M_2(-1)} \begin{pmatrix} 1 & -\frac{3}{2} & | & \frac{1}{2} & 0 \\ 0 & 1 & | & -2 & -1 \end{pmatrix}$$

$$\xrightarrow{A_{2,1}(\frac{3}{2})} \begin{pmatrix} 1 & 0 & | & -\frac{5}{2} & -\frac{3}{2} \\ 0 & 1 & | & -2 & -1 \end{pmatrix}.$$

Thus $x = -\frac{5}{2}$, $y = -\frac{3}{2}$, $z = -2$, $w = -1$, and $A^{-1} = \begin{pmatrix} -\frac{5}{2} & -\frac{3}{2} \\ -2 & -1 \end{pmatrix}$. We still must check our answer. We have

$$AA^{-1} = \begin{pmatrix} 2 & -3 \\ -4 & 5 \end{pmatrix} \begin{pmatrix} -\frac{5}{2} & -\frac{3}{2} \\ -2 & -1 \end{pmatrix} = \begin{pmatrix} 1 & 0 \\ 0 & 1 \end{pmatrix}$$

and

$$A^{-1}A = \begin{pmatrix} -\frac{5}{2} & -\frac{3}{2} \\ -2 & -1 \end{pmatrix} \begin{pmatrix} 2 & -3 \\ -4 & 5 \end{pmatrix} = \begin{pmatrix} 1 & 0 \\ 0 & 1 \end{pmatrix}.$$

Thus A is invertible and $A^{-1} = \begin{pmatrix} -\frac{5}{2} & -\frac{3}{2} \\ -2 & -1 \end{pmatrix}$.

EXAMPLE 3 Let $A = \begin{pmatrix} 1 & 2 \\ -2 & -4 \end{pmatrix}$. Calculate A^{-1} if it exists.

Solution If $A^{-1} = \begin{pmatrix} x & y \\ z & w \end{pmatrix}$ exists, then

$$AA^{-1} = \begin{pmatrix} 1 & 2 \\ -2 & -4 \end{pmatrix} \begin{pmatrix} x & y \\ z & w \end{pmatrix} = \begin{pmatrix} x + 2z & y + 2w \\ -2x - 4z & -2y - 4w \end{pmatrix} = \begin{pmatrix} 1 & 0 \\ 0 & 1 \end{pmatrix}.$$

This leads to the system

$$\begin{aligned} x + 2z &= 1 \\ y + 2w &= 0 \\ -2x - 4z &= 0 \\ -2y - 4w &= 1. \end{aligned} \qquad (10)$$

Using the same reasoning as in Example 2, we can write this system in the augmented matrix form $(A|I)$ and row-reduce.

$$\begin{pmatrix} 1 & 2 & | & 1 & 0 \\ -2 & -4 & | & 0 & 1 \end{pmatrix} \xrightarrow{A_{1,2}(2)} \begin{pmatrix} 1 & 2 & | & 1 & 0 \\ 0 & 0 & | & 2 & 1 \end{pmatrix}$$

This is as far as we can go. The last line reads $0 = 2$ or $0 = 1$, depending on which of the two systems of equations (in x and z or in y and w) is being solved. Thus system (10) is inconsistent and A is not invertible.

The last two examples illustrate a procedure that always works when you are trying to find the inverse of a matrix.

PROCEDURE FOR COMPUTING THE INVERSE OF A SQUARE MATRIX A

Step 1. Write the augmented matrix $(A|I)$.

Step 2. Use row reduction to reduce the matrix A to its reduced row echelon form.

Step 3. Decide if A is invertible.

(a) If A can be reduced to the identity matrix I, then A^{-1} will be the matrix to the right of the vertical bar.

(b) If the row reduction of A leads to a row of zeros to the left of the vertical bar, then A is not invertible.

Remark We can rephrase (a) and (b) as follows.

A square matrix A is invertible if and only if its reduced row echelon form is the identity matrix.

2.7/THE INVERSE OF A SQUARE MATRIX

Let $A = \begin{pmatrix} a_{11} & a_{12} \\ a_{21} & a_{22} \end{pmatrix}$. Then, as in Equation (1.2.7, page 4), we define

$$\text{Determinant of } A = a_{11}a_{22} - a_{12}a_{21} \quad (11)$$

We abbreviate the determinant of A by $\det A$.

THEOREM 4   Let A be a 2×2 matrix. Then:

i. A is invertible if and only if $\det A \neq 0$.
ii. If $\det A \neq 0$, then

$$A^{-1} = \frac{1}{\det A} \begin{pmatrix} a_{22} & -a_{12} \\ -a_{21} & a_{11} \end{pmatrix}^{\dagger} \quad (12)$$

Proof   First suppose that $\det A \neq 0$ and let $B = (1/\det A)\begin{pmatrix} a_{22} & -a_{12} \\ -a_{21} & a_{11} \end{pmatrix}$. Then

$$BA = \frac{1}{\det A} \begin{pmatrix} a_{22} & -a_{12} \\ -a_{21} & a_{11} \end{pmatrix} \begin{pmatrix} a_{11} & a_{12} \\ a_{21} & a_{22} \end{pmatrix}$$

$$= \frac{1}{a_{11}a_{22} - a_{12}a_{21}} \begin{pmatrix} a_{22}a_{11} - a_{12}a_{21} & 0 \\ 0 & -a_{21}a_{12} + a_{11}a_{22} \end{pmatrix} = \begin{pmatrix} 1 & 0 \\ 0 & 1 \end{pmatrix} = I$$

Similarly $AB = I$, which shows that A is invertible and that $B = A^{-1}$. We still must show that if A is invertible, then $\det A \neq 0$. To do so, we consider the system

$$\begin{aligned} a_{11}x_1 + a_{12}x_2 &= b_1 \\ a_{21}x_1 + a_{22}x_2 &= b_2 \end{aligned} \quad (13)$$

We do this because we know from our Summing Up Theorem (Theorem 1.2.1, page 4) that if this system has a unique solution, then its determinant is nonzero. The system can be written in the form

$$A\mathbf{x} = \mathbf{b} \quad (14)$$

with $\mathbf{x} = \begin{pmatrix} x_1 \\ x_2 \end{pmatrix}$ and $\mathbf{b} = \begin{pmatrix} b_1 \\ b_2 \end{pmatrix}$. Then, since A is invertible, we see from (2) that the system (14) has a unique solution given by

$$\mathbf{x} = A^{-1}\mathbf{b}$$

But by Theorem 1.2.1, the fact that system (13) has a unique solution implies that $\det A \neq 0$. This completes the proof. ∎

---

† This formula can be obtained directly by applying our procedure for computing an inverse (see Problem 46).

EXAMPLE 4 Let $A = \begin{pmatrix} 2 & -4 \\ 1 & 3 \end{pmatrix}$. Calculate A^{-1} if it exists.

Solution We find that $\det A = (2)(3) - (-4)(1) = 10$; hence A^{-1} exists. From Equation (12), we get

$$A^{-1} = \frac{1}{10}\begin{pmatrix} 3 & 4 \\ -1 & 2 \end{pmatrix} = \begin{pmatrix} \frac{3}{10} & \frac{4}{10} \\ -\frac{1}{10} & \frac{2}{10} \end{pmatrix}.$$

Check.

$$A^{-1}A = \frac{1}{10}\begin{pmatrix} 3 & 4 \\ -1 & 2 \end{pmatrix}\begin{pmatrix} 2 & -4 \\ 1 & 3 \end{pmatrix} = \frac{1}{10}\begin{pmatrix} 10 & 0 \\ 0 & 10 \end{pmatrix} = \begin{pmatrix} 1 & 0 \\ 0 & 1 \end{pmatrix}$$

and

$$AA^{-1} = \begin{pmatrix} 2 & -4 \\ 1 & 3 \end{pmatrix}\begin{pmatrix} \frac{3}{10} & \frac{4}{10} \\ -\frac{1}{10} & \frac{2}{10} \end{pmatrix} = \begin{pmatrix} 1 & 0 \\ 0 & 1 \end{pmatrix}.$$

EXAMPLE 5 Let $A = \begin{pmatrix} 1 & 2 \\ -2 & -4 \end{pmatrix}$. Calculate A^{-1} if it exists.

Solution We find that $\det A = (1)(-4) - (2)(-2) = -4 + 4 = 0$, so that A^{-1} does not exist, as we saw in Example 3.

The procedure described above works for $n \times n$ matrices where $n > 2$. We illustrate this with a number of examples.

EXAMPLE 6 Let $A = \begin{pmatrix} 2 & 4 & 6 \\ 4 & 5 & 6 \\ 3 & 1 & -2 \end{pmatrix}$ (see Example 2.5.1 on page 50). Calculate A^{-1} if it exists.

Solution We first put I next to A in an augmented matrix form

$$\begin{pmatrix} 2 & 4 & 6 & | & 1 & 0 & 0 \\ 4 & 5 & 6 & | & 0 & 1 & 0 \\ 3 & 1 & -2 & | & 0 & 0 & 1 \end{pmatrix}$$

and then carry out the row reduction.

$$\xrightarrow{M_1(\frac{1}{2})} \begin{pmatrix} 1 & 2 & 3 & | & \frac{1}{2} & 0 & 0 \\ 4 & 5 & 6 & | & 0 & 1 & 0 \\ 3 & 1 & -2 & | & 0 & 0 & 1 \end{pmatrix} \xrightarrow[A_{1,3}(-3)]{A_{1,2}(-4)} \begin{pmatrix} 1 & 2 & 3 & | & \frac{1}{2} & 0 & 0 \\ 0 & -3 & -6 & | & -2 & 1 & 0 \\ 0 & -5 & -11 & | & -\frac{3}{2} & 0 & 1 \end{pmatrix}$$

$$\xrightarrow{M_2(-\frac{1}{3})} \begin{pmatrix} 1 & 2 & 3 & | & \frac{1}{2} & 0 & 0 \\ 0 & 1 & 2 & | & \frac{2}{3} & -\frac{1}{3} & 0 \\ 0 & -5 & -11 & | & -\frac{3}{2} & 0 & 1 \end{pmatrix} \xrightarrow[A_{2,3}(5)]{A_{2,1}(-2)} \begin{pmatrix} 1 & 0 & -1 & | & -\frac{5}{6} & \frac{2}{3} & 0 \\ 0 & 1 & 2 & | & \frac{2}{3} & -\frac{1}{3} & 0 \\ 0 & 0 & -1 & | & \frac{11}{6} & -\frac{5}{3} & 1 \end{pmatrix}$$

$$\xrightarrow{M_3(-1)} \begin{pmatrix} 1 & 0 & -1 & | & -\frac{5}{6} & \frac{2}{3} & 0 \\ 0 & 1 & 2 & | & \frac{2}{3} & -\frac{1}{3} & 0 \\ 0 & 0 & 1 & | & -\frac{11}{6} & \frac{5}{3} & -1 \end{pmatrix} \xrightarrow[A_{3,2}(-2)]{A_{3,1}(1)} \begin{pmatrix} 1 & 0 & 0 & | & -\frac{8}{3} & \frac{7}{3} & -1 \\ 0 & 1 & 0 & | & \frac{13}{3} & -\frac{11}{3} & 2 \\ 0 & 0 & 1 & | & -\frac{11}{6} & \frac{5}{3} & -1 \end{pmatrix}$$

Since A has now been reduced to I, we have

$$A^{-1} = \begin{pmatrix} -\frac{8}{3} & \frac{7}{3} & -1 \\ \frac{13}{3} & -\frac{11}{3} & 2 \\ -\frac{11}{6} & \frac{5}{3} & -1 \end{pmatrix} = \frac{1}{6}\begin{pmatrix} -16 & 14 & -6 \\ 26 & -22 & 12 \\ -11 & 10 & -6 \end{pmatrix} \quad \begin{array}{l} \text{We factor out} \\ \frac{1}{6} \text{ to make} \\ \text{computations} \\ \text{easier.} \end{array}$$

Check. $A^{-1}A = \frac{1}{6}\begin{pmatrix} -16 & 14 & -6 \\ 26 & -22 & 12 \\ -11 & 10 & -6 \end{pmatrix}\begin{pmatrix} 2 & 4 & 6 \\ 4 & 5 & 6 \\ 3 & 1 & -2 \end{pmatrix} = \frac{1}{6}\begin{pmatrix} 6 & 0 & 0 \\ 0 & 6 & 0 \\ 0 & 0 & 6 \end{pmatrix} = I.$

We can also verify that $AA^{-1} = I$.

Warning. It is easy to make numerical errors in computing A^{-1}. Therefore it is essential to check the computations by verifying that $A^{-1}A = I$.

EXAMPLE 7 Let $A = \begin{pmatrix} 2 & 4 & 3 \\ 0 & 1 & -1 \\ 3 & 5 & 7 \end{pmatrix}$. Calculate A^{-1} if it exists.

Solution Proceeding as in Example 6 we obtain, successively, the following augmented matrices:

$$\begin{pmatrix} 2 & 4 & 3 & | & 1 & 0 & 0 \\ 0 & 1 & -1 & | & 0 & 1 & 0 \\ 3 & 5 & 7 & | & 0 & 0 & 1 \end{pmatrix} \xrightarrow{M_1(\frac{1}{2})} \begin{pmatrix} 1 & 2 & \frac{3}{2} & | & \frac{1}{2} & 0 & 0 \\ 0 & 1 & -1 & | & 0 & 1 & 0 \\ 3 & 5 & 7 & | & 0 & 0 & 1 \end{pmatrix}$$

$$\xrightarrow{A_{1,3}(-3)} \begin{pmatrix} 1 & 2 & \frac{3}{2} & | & \frac{1}{2} & 0 & 0 \\ 0 & 1 & -1 & | & 0 & 1 & 0 \\ 0 & -1 & \frac{5}{2} & | & -\frac{3}{2} & 0 & 1 \end{pmatrix} \xrightarrow[A_{2,3}(1)]{A_{2,1}(-2)} \begin{pmatrix} 1 & 0 & \frac{7}{2} & | & \frac{1}{2} & -2 & 0 \\ 0 & 1 & -1 & | & 0 & 1 & 0 \\ 0 & 0 & \frac{3}{2} & | & -\frac{3}{2} & 1 & 1 \end{pmatrix}$$

$$\xrightarrow{M_3(\frac{2}{3})} \begin{pmatrix} 1 & 0 & \frac{7}{2} & | & \frac{1}{2} & -2 & 0 \\ 0 & 1 & -1 & | & 0 & 1 & 0 \\ 0 & 0 & 1 & | & -1 & \frac{2}{3} & \frac{2}{3} \end{pmatrix} \xrightarrow[A_{3,2}(1)]{A_{3,1}(-\frac{7}{2})} \begin{pmatrix} 1 & 0 & 0 & | & 4 & -\frac{13}{3} & -\frac{7}{3} \\ 0 & 1 & 0 & | & -1 & \frac{5}{3} & \frac{2}{3} \\ 0 & 0 & 1 & | & -1 & \frac{2}{3} & \frac{2}{3} \end{pmatrix}$$

Thus

$$A^{-1} = \begin{pmatrix} 4 & -\frac{13}{3} & -\frac{7}{3} \\ -1 & \frac{5}{3} & \frac{2}{3} \\ -1 & \frac{2}{3} & \frac{2}{3} \end{pmatrix}$$

Check. $A^{-1}A = \begin{pmatrix} 4 & -\frac{13}{3} & -\frac{7}{3} \\ -1 & \frac{5}{3} & \frac{2}{3} \\ -1 & \frac{2}{3} & \frac{2}{3} \end{pmatrix} \begin{pmatrix} 2 & 4 & 3 \\ 0 & 1 & -1 \\ 3 & 5 & 7 \end{pmatrix} = \begin{pmatrix} 1 & 0 & 0 \\ 0 & 1 & 0 \\ 0 & 0 & 1 \end{pmatrix}$

EXAMPLE 8 Let $A = \begin{pmatrix} 1 & -3 & 4 \\ 2 & -5 & 7 \\ 0 & -1 & 1 \end{pmatrix}$. Calculate A^{-1} if it exists.

Solution Proceeding as before we obtain, successively,

$$\begin{pmatrix} 1 & -3 & 4 & | & 1 & 0 & 0 \\ 2 & -5 & 7 & | & 0 & 1 & 0 \\ 0 & -1 & 1 & | & 0 & 0 & 1 \end{pmatrix} \xrightarrow{A_{1,2}(-2)} \begin{pmatrix} 1 & -3 & 4 & | & 1 & 0 & 0 \\ 0 & 1 & -1 & | & -2 & 1 & 0 \\ 0 & -1 & 1 & | & 0 & 0 & 1 \end{pmatrix}$$

$$\xrightarrow{\substack{A_{2,1}(3) \\ A_{2,3}(1)}} \begin{pmatrix} 1 & 0 & 1 & | & -5 & 3 & 0 \\ 0 & 1 & -1 & | & -2 & 1 & 0 \\ 0 & 0 & 0 & | & -2 & 1 & 1 \end{pmatrix}$$

This is as far as we can go. The matrix A *cannot* be reduced to the identity matrix and we can conclude that A is *not* invertible.

There is another way to see the result of the last example. Let **b** be any 3-vector and consider the system $A\mathbf{x} = \mathbf{b}$. If we tried to solve this by Gaussian elimination, we would end up with an equation that reads $0 = c \neq 0$ as in Example 3, or $0 = 0$. This is case (*ii*) or (*iii*) of Section 1.3 (see page 16). That is, the system either has no solution or it has an infinite number of solutions. The one possibility ruled out is the case in which the system has a unique solution. But if A^{-1} existed, then there would be a unique solution given by $\mathbf{x} = A^{-1}\mathbf{b}$. We are left to conclude that:

> If in the row reduction of A we end up with a row of zeros, then A is *not* invertible.

DEFINITION 3 **ROW EQUIVALENT MATRICES** Suppose that by elementary row operations we can transform the matrix A into the matrix B. Then A and B are said to be **row equivalent**.

The reasoning used above can be used to prove the following theorem (see Problem 47).

THEOREM 5   Let A be an $n \times n$ matrix.

i. A is invertible if and only if A is row equivalent to the identity matrix I_n; that is, the reduced row echelon form of A is I_n.
ii. A is invertible if and only if the system $A\mathbf{x} = \mathbf{b}$ has a unique solution for every n-vector $\mathbf{b}$.
iii. If A is invertible, then this unique solution is given by $\mathbf{x} = A^{-1}\mathbf{b}$.

EXAMPLE 9   Solve the system

$$2x_1 + 4x_2 + 3x_3 = 6$$
$$x_2 - x_3 = -4$$
$$3x_1 + 5x_2 + 7x_3 = 7$$

Solution   This system can be written as $A\mathbf{x} = \mathbf{b}$, where $A = \begin{pmatrix} 2 & 4 & 3 \\ 0 & 1 & -1 \\ 3 & 5 & 7 \end{pmatrix}$ and $\mathbf{b} = \begin{pmatrix} 6 \\ -4 \\ 7 \end{pmatrix}$. In Example 7 we found that A^{-1} exists and

$$A^{-1} = \begin{pmatrix} 4 & -\frac{13}{3} & -\frac{7}{3} \\ -1 & \frac{5}{3} & \frac{2}{3} \\ -1 & \frac{2}{3} & \frac{2}{3} \end{pmatrix}$$

Thus the unique solution is given by

$$\mathbf{x} = \begin{pmatrix} x_1 \\ x_2 \\ x_3 \end{pmatrix} = A^{-1}\mathbf{b} = \begin{pmatrix} 4 & -\frac{13}{3} & -\frac{7}{3} \\ -1 & \frac{5}{3} & \frac{2}{3} \\ -1 & \frac{2}{3} & \frac{2}{3} \end{pmatrix} \begin{pmatrix} 6 \\ -4 \\ 7 \end{pmatrix} = \begin{pmatrix} 25 \\ -8 \\ -4 \end{pmatrix}$$

EXAMPLE 10   In the Leontief input-output model described in Example 1.3.8 on page 17, we obtained the system

$$a_{11}x_1 + a_{12}x_2 + \cdots + a_{1n}x_n + e_1 = x_1$$
$$a_{21}x_1 + a_{22}x_2 + \cdots + a_{2n}x_n + e_2 = x_2$$
$$\vdots \qquad \vdots \qquad \qquad \vdots \qquad \vdots \quad \vdots \qquad (15)$$
$$a_{n1}x_1 + a_{n2}x_2 + \cdots + a_{nn}x_n + e_n = x_n$$

which can be written as

$$A\mathbf{x} + \mathbf{e} = \mathbf{x} = I\mathbf{x}$$

or

$$(I - A)\mathbf{x} = \mathbf{e} \qquad (16)$$

The matrix A of internal demands is called the **technology matrix**, and the matrix $I - A$ is called the **Leontief matrix**. If the Leontief matrix is invertible, then systems (15) and (16) have unique solutions.

Leontief used his model to analyze the 1958 American economy.† He divided the economy into 81 sectors and grouped them into six families of related sectors. For simplicity, we treat each family of sectors as a single sector so we can treat the American economy as an economy with six industries. These industries are listed in Table 1.

Table 1

Sector	Examples
Final nonmetal (FN)	Furniture, processed food
Final metal (FM)	Household appliances, motor vehicles
Basic metal (BM)	Machine-shop products, mining
Basic nonmetal (BN)	Agriculture, printing
Energy (E)	Petroleum, coal
Services (S)	Amusements, real estate

The input-output table, Table 2, gives internal demands in 1958 based on Leontief's figures. The units in the table are millions of dollars. Thus, for example, the number 0.173 in the 6,5 position means that in order to produce $1 million worth of energy, it is necessary to provide $0.173 million = $173,000 worth of services. Similarly, the 0.037 in the 4,2 position means that in order to produce $1 million worth of final metal, it is necessary to expend $0.037 million = $37,000 on basic nonmetal products.

Table 2 Internal Demands in 1958 U.S. Economy

	FN	FM	BM	BN	E	S
FN	0.170	0.004	0	0.029	0	0.008
FM	0.003	0.295	0.018	0.002	0.004	0.016
BM	0.025	0.173	0.460	0.007	0.011	0.007
BN	0.348	0.037	0.021	0.403	0.011	0.048
E	0.007	0.001	0.039	0.025	0.358	0.025
S	0.120	0.074	0.104	0.123	0.173	0.234

Finally, Leontief estimated the following external demands on the 1958 American economy (in millions of dollars).

Table 3 External Demands on 1958 U.S. Economy (Millions of Dollars)

FN	$99,640
FM	$75,548
BM	$14,444
BN	$33,501
E	$23,527
S	$263,985

†*Scientific American* (April 1965), pp. 26–27.

In order to run the American economy in 1958 and meet all external demands, how many units in each of the six sectors had to be produced?

Solution The technology matrix is given by

$$A = \begin{pmatrix} 0.170 & 0.004 & 0 & 0.029 & 0 & 0.008 \\ 0.003 & 0.295 & 0.018 & 0.002 & 0.004 & 0.016 \\ 0.025 & 0.173 & 0.460 & 0.007 & 0.011 & 0.007 \\ 0.348 & 0.037 & 0.021 & 0.403 & 0.011 & 0.048 \\ 0.007 & 0.001 & 0.039 & 0.025 & 0.358 & 0.025 \\ 0.120 & 0.074 & 0.104 & 0.123 & 0.173 & 0.234 \end{pmatrix}$$

and

$$\mathbf{e} = \begin{pmatrix} 99{,}640 \\ 75{,}548 \\ 14{,}444 \\ 33{,}501 \\ 23{,}527 \\ 263{,}985 \end{pmatrix}$$

To obtain the Leontief matrix, we subtract to obtain

$$I - A = \begin{pmatrix} 1 & 0 & 0 & 0 & 0 & 0 \\ 0 & 1 & 0 & 0 & 0 & 0 \\ 0 & 0 & 1 & 0 & 0 & 0 \\ 0 & 0 & 0 & 1 & 0 & 0 \\ 0 & 0 & 0 & 0 & 1 & 0 \\ 0 & 0 & 0 & 0 & 0 & 1 \end{pmatrix}$$

$$- \begin{pmatrix} 0.170 & 0.004 & 0 & 0.029 & 0 & 0.008 \\ 0.003 & 0.295 & 0.018 & 0.002 & 0.004 & 0.016 \\ 0.025 & 0.173 & 0.460 & 0.007 & 0.011 & 0.007 \\ 0.348 & 0.037 & 0.021 & 0.403 & 0.011 & 0.048 \\ 0.007 & 0.001 & 0.039 & 0.025 & 0.358 & 0.025 \\ 0.120 & 0.074 & 0.104 & 0.123 & 0.173 & 0.234 \end{pmatrix}$$

$$= \begin{pmatrix} 0.830 & -0.004 & 0 & -0.029 & 0 & -0.008 \\ -0.003 & 0.705 & -0.018 & -0.002 & -0.004 & -0.016 \\ -0.025 & -0.173 & 0.540 & -0.007 & -0.011 & -0.007 \\ -0.348 & -0.037 & -0.021 & 0.597 & -0.011 & -0.048 \\ -0.007 & -0.001 & -0.039 & -0.025 & 0.642 & -0.025 \\ -0.120 & -0.074 & -0.104 & -0.123 & -0.173 & 0.766 \end{pmatrix}$$

The computation of the inverse of a 6 × 6 matrix is a tedious affair. Carrying three decimal places on a calculator, we obtain the matrix below. Intermediate steps are omitted.

$$(I - A)^{-1} = \begin{pmatrix} 1.234 & 0.014 & 0.006 & 0.064 & 0.007 & 0.018 \\ 0.017 & 1.436 & 0.057 & 0.012 & 0.020 & 0.032 \\ 0.071 & 0.465 & 1.877 & 0.019 & 0.045 & 0.031 \\ 0.751 & 0.134 & 0.100 & 1.740 & 0.066 & 0.124 \\ 0.060 & 0.045 & 0.130 & 0.082 & 1.578 & 0.059 \\ 0.339 & 0.236 & 0.307 & 0.312 & 0.376 & 1.349 \end{pmatrix}$$

Therefore the "ideal" output vector is given by

$$\mathbf{x} = (I - A)^{-1}\mathbf{e} = \begin{pmatrix} 1.234 & 0.014 & 0.006 & 0.064 & 0.007 & 0.018 \\ 0.017 & 1.436 & 0.057 & 0.012 & 0.020 & 0.032 \\ 0.071 & 0.465 & 1.877 & 0.019 & 0.045 & 0.031 \\ 0.751 & 0.134 & 0.100 & 1.740 & 0.066 & 0.124 \\ 0.060 & 0.045 & 0.130 & 0.082 & 1.578 & 0.059 \\ 0.339 & 0.236 & 0.307 & 0.312 & 0.376 & 1.349 \end{pmatrix} \begin{pmatrix} 99{,}640 \\ 75{,}548 \\ 14{,}444 \\ 33{,}501 \\ 23{,}527 \\ 263{,}985 \end{pmatrix}$$

$$= \begin{pmatrix} 131{,}161 \\ 120{,}324 \\ 79{,}194 \\ 178{,}936 \\ 66{,}703 \\ 426{,}542 \end{pmatrix}$$

This means that it would require 131,161 units ($131,161 million worth) of final nonmetal products, 120,324 units of final metal products, 79,194 units of basic metal products, 178,936 units of basic nonmetal products, 66,703 units of energy and 426,542 service units to run the U.S. economy and meet the external demands in 1958.

In Section 1.2 we encountered the first form of our Summing Up Theorem (Theorem 1.2.1, page 4). We are now ready to improve upon it. The next theorem states that several statements involving inverse, uniqueness of solutions, row equivalence, linear independence, and determinants are equivalent. This means that if one of the statements is true, all the statements are true. At this point we can prove the equivalence of parts (*i*), (*ii*), (*iii*), and (*iv*). We shall finish the proof after we have developed some basic theory about determinants (see Theorem 3.4.4 on page 110).

THEOREM 6   **SUMMING UP THEOREM—VIEW 2**   Let A be an $n \times n$ matrix. Then each of the following six statements implies the other five (that is, if one is true, all are true):

i. A is invertible.
ii. The only solution to the homogeneous system $A\mathbf{x} = \mathbf{0}$ is the trivial solution ($\mathbf{x} = \mathbf{0}$).

iii. The system $A\mathbf{x} = \mathbf{b}$ has a unique solution for every n-vector $\mathbf{b}$.
iv. A is row equivalent to the $n \times n$ identity matrix I_n.
v. The rows (and columns) of A are linearly independent.
vi. $\det A \neq 0$. (So far, $\det A$ is only defined if A is a 2×2 matrix.)

Proof We have already seen that statements (*i*) and (*iii*) are equivalent (Theorem 5 (part *ii*)) and that (*i*) and (*iv*) are equivalent (Theorem 5 (part *i*)). We shall see that (*ii*) and (*iv*) are equivalent. Suppose that (*ii*) holds. That is, suppose that $A\mathbf{x} = \mathbf{0}$ has only the trivial solution $\mathbf{x} = \mathbf{0}$. If we write out this system we obtain

$$\begin{aligned} a_{11}x_1 + a_{12}x_2 + \cdots + a_{1n}x_n &= 0 \\ a_{21}x_1 + a_{22}x_2 + \cdots + a_{2n}x_n &= 0 \\ &\vdots \\ a_{n1}x_1 + a_{n2}x_2 + \cdots + a_{nn}x_n &= 0 \end{aligned} \qquad (17)$$

If A were not equivalent to I_n, then row reduction of the augmented matrix associated with (17) would leave us with a row of zeros. But if, say, the last row is zero, then the last equation reads $0 = 0$. Then, the homogeneous system reduces to one with $n - 1$ equations in n unknowns which, by Theorem 1.4.1 on page 24 has an infinite number of solutions. But we assumed that $\mathbf{x} = \mathbf{0}$ was the only solution to system (17). This contradiction shows that A is row equivalent to I_n. Conversely, suppose that (*iv*) holds; that is, suppose that A is row equivalent to I_n. Then by Theorem 5 (part *i*), A is invertible and by Theorem 5 (part *iii*) the unique solution to $A\mathbf{x} = \mathbf{0}$ is $\mathbf{x} = A^{-1}\mathbf{0} = \mathbf{0}$. Thus (*ii*) and (*iv*) are equivalent. In Theorem 1.2.1 we showed that (*i*) and (*vi*) are equivalent in the 2×2 case. It is not difficult to prove the equivalence of (*v*) and (*vi*) in the 2×2 case (see Problem 38). We shall prove the equivalence of (*ii*), (*v*), and (*vi*) in Section 3.4. ∎

Remark. We could add another statement to the theorem. Suppose the system $A\mathbf{x} = \mathbf{b}$ has a unique solution. Let R be a matrix in row echelon form that is row equivalent to A. Then R cannot have a row of zeros because if it did, it could not be reduced to the identity matrix.† Thus the row echelon form of A must look like this:

$$\begin{pmatrix} 1 & r_{12} & r_{13} & \cdots & r_{1n} \\ 0 & 1 & r_{23} & \cdots & r_{2n} \\ 0 & 0 & 1 & \cdots & r_{3n} \\ \vdots & \vdots & \vdots & & \vdots \\ 0 & 0 & 0 & \cdots & 1 \end{pmatrix} \qquad (18)$$

That is, R is a matrix with 1's down the diagonal and 0's below it. We thus have Theorem 7.

---

† Note that if the *i*th row of R contains only zeros, then the homogeneous system $R\mathbf{x} = \mathbf{0}$ contains more unknowns than equations (since the *i*th equation is the zero equation) and the system has an infinite number of solutions. But then $A\mathbf{x} = \mathbf{0}$ has an infinite number of solutions, which is a contradiction of our assumption.

THEOREM 7 If any of the statements in Theorem 6 holds, then the row echelon form of A has the form of matrix (18).

We have seen that in order to verify that $B = A^{-1}$, we have to check that $AB = BA = I$. It turns out that only half this work has to be done.

THEOREM 8 Let A and B be $n \times n$ matrices. Then A is invertible and $B = A^{-1}$ if (i) $BA = I$ or (ii) $AB = I$.

Remark. This theorem simplifies the work in checking that one matrix is the inverse of another.

Proof **i.** We assume that $BA = I$. Consider the homogeneous system $A\mathbf{x} = \mathbf{0}$. Multiplying both sides of this equation on the left by B, we obtain

$$BA\mathbf{x} = B\mathbf{0} \tag{19}$$

But $BA = I$ and $B\mathbf{0} = \mathbf{0}$, so (19) becomes $I\mathbf{x} = \mathbf{0}$ or $\mathbf{x} = \mathbf{0}$. This shows that $\mathbf{x} = \mathbf{0}$ is the only solution to $A\mathbf{x} = \mathbf{0}$ and, by Theorem 6 (parts *i* and *ii*), this means that A is invertible. We still have to show that $B = A^{-1}$. Let $A^{-1} = C$. Then $AC = I$. Thus $BAC = B(AC) = BI = B$ and $BAC = (BA)C = IC = C$. Hence $B = C$ and part (*i*) is proved.

ii. Let $AB = I$. Then, from part (*i*), $A = B^{-1}$. From Definition 2 this means that $AB = BA = I$, which proves that A is invertible and that $B = A^{-1}$. This completes the proof. ∎

PROBLEMS 2.7 In Problems 1–15 determine whether the given matrix is invertible. If it is, calculate the inverse.

1. $\begin{pmatrix} 2 & 1 \\ 3 & 2 \end{pmatrix}$
2. $\begin{pmatrix} -1 & 6 \\ 2 & -12 \end{pmatrix}$
3. $\begin{pmatrix} 0 & 1 \\ 1 & 0 \end{pmatrix}$
4. $\begin{pmatrix} 1 & 1 \\ 3 & 3 \end{pmatrix}$
5. $\begin{pmatrix} a & a \\ b & b \end{pmatrix}$
6. $\begin{pmatrix} 1 & 1 & 1 \\ 0 & 2 & 3 \\ 5 & 5 & 1 \end{pmatrix}$
7. $\begin{pmatrix} 3 & 2 & 1 \\ 0 & 2 & 2 \\ 0 & 0 & -1 \end{pmatrix}$
8. $\begin{pmatrix} 1 & 1 & 1 \\ 0 & 1 & 1 \\ 0 & 0 & 1 \end{pmatrix}$
9. $\begin{pmatrix} 1 & 6 & 2 \\ -2 & 3 & 5 \\ 7 & 12 & -4 \end{pmatrix}$
10. $\begin{pmatrix} 3 & 1 & 0 \\ 1 & -1 & 2 \\ 1 & 1 & 1 \end{pmatrix}$
11. $\begin{pmatrix} 2 & -1 & 4 \\ -1 & 0 & 5 \\ 19 & -7 & 3 \end{pmatrix}$
12. $\begin{pmatrix} 1 & 2 & 3 \\ 1 & 1 & 2 \\ 0 & 1 & 2 \end{pmatrix}$
13. $\begin{pmatrix} 1 & 1 & 1 & 1 \\ 1 & 2 & -1 & 2 \\ 1 & -1 & 2 & 1 \\ 1 & 3 & 3 & 2 \end{pmatrix}$
14. $\begin{pmatrix} 1 & 0 & 2 & 3 \\ -1 & 1 & 0 & 4 \\ 2 & 1 & -1 & 3 \\ -1 & 0 & 5 & 7 \end{pmatrix}$
15. $\begin{pmatrix} 1 & -3 & 0 & -2 \\ 3 & -12 & -2 & -6 \\ -2 & 10 & 2 & 5 \\ -1 & 6 & 1 & 3 \end{pmatrix}$

16. Show that if A, B, and C are invertible matrices, then ABC is invertible and $(ABC)^{-1} = C^{-1}B^{-1}A^{-1}$.

17. If $A_1, A_2, \ldots, A_m$ are invertible $n \times n$ matrices, show that $A_1 A_2 \cdots A_m$ is invertible and calculate its inverse.

18. Show that the matrix $\begin{pmatrix} 3 & 4 \\ -2 & -3 \end{pmatrix}$ is equal to its own inverse.

19. Show that the matrix $\begin{pmatrix} a_{11} & a_{12} \\ a_{21} & a_{22} \end{pmatrix}$ is equal to its own inverse if $A = \pm I$ or if $a_{11} = -a_{22}$ and $a_{21} a_{12} = 1 - a_{11}^2$.

20. Find the output vector $\mathbf{x}$ in the Leontief input–output model if $n = 3$, $\mathbf{e} = \begin{pmatrix} 30 \\ 20 \\ 40 \end{pmatrix}$, and $A = \begin{pmatrix} \frac{1}{5} & \frac{1}{5} & 0 \\ \frac{2}{5} & \frac{2}{5} & \frac{3}{5} \\ \frac{1}{5} & \frac{1}{10} & \frac{2}{5} \end{pmatrix}$.

*21. Suppose that A is $n \times m$ and B is $m \times n$ so that AB is $n \times n$. Show that AB is not invertible if $n > m$. [*Hint:* Show that there is a nonzero vector $\mathbf{x}$ such that $AB\mathbf{x} = \mathbf{0}$ and then apply Theorem 6.]

*22. Use the methods of this section to find the inverses of the following matrices with complex entries:

a. $\begin{pmatrix} i & 2 \\ 1 & -i \end{pmatrix}$ **b.** $\begin{pmatrix} 1-i & 0 \\ 0 & 1+i \end{pmatrix}$ **c.** $\begin{pmatrix} 1 & i & 0 \\ -i & 0 & 1 \\ 0 & 1+i & 1-i \end{pmatrix}$

23. Show that for every real number θ the matrix $\begin{pmatrix} \sin \theta & \cos \theta & 0 \\ \cos \theta & -\sin \theta & 0 \\ 0 & 0 & 1 \end{pmatrix}$ is invertible and find its inverse.

24. Calculate the inverse of $A = \begin{pmatrix} 2 & 0 & 0 \\ 0 & 3 & 0 \\ 0 & 0 & 4 \end{pmatrix}$.

25. A square matrix $A = (a_{ij})$ is called **diagonal** if all its elements off the main diagonal are zero. That is, $a_{ij} = 0$ if $i \neq j$. (The matrix of Problem 24 is diagonal.) Show that a diagonal matrix is invertible if and only if each of its diagonal components is nonzero.

26. Let

$$A = \begin{pmatrix} a_{11} & 0 & \cdots & 0 \\ 0 & a_{22} & \cdots & 0 \\ \vdots & & \ddots & \\ 0 & 0 & \cdots & a_{nn} \end{pmatrix}$$

be a diagonal matrix such that each of its diagonal components is nonzero. Calculate A^{-1}.

27. Calculate the inverse of $A = \begin{pmatrix} 2 & 1 & -1 \\ 0 & 3 & 4 \\ 0 & 0 & 5 \end{pmatrix}$.

28. Show that the matrix $A = \begin{pmatrix} 1 & 0 & 0 \\ -2 & 0 & 0 \\ 4 & 6 & 1 \end{pmatrix}$ is not invertible.

*29. A square matrix is called **upper (lower) triangular** if all its elements below (above) the main diagonal are zero. (The matrix of Problem 27 is upper triangular and the matrix of Problem 28 is lower triangular.) Show that an upper or lower triangular matrix is invertible if and only if each of its diagonal elements is nonzero.

*30. Show that the inverse of an invertible upper triangular matrix is upper triangular. [*Hint:* First prove the result for a 3×3 matrix.]

In Problems 31 and 32 a matrix is given. In each case show that the matrix is not invertible by finding a nonzero vector $\mathbf{x}$ such that $A\mathbf{x} = \mathbf{0}$.

31. $\begin{pmatrix} 2 & -1 \\ -4 & 2 \end{pmatrix}$

32. $\begin{pmatrix} 1 & -1 & 3 \\ 0 & 4 & -2 \\ 2 & -6 & 8 \end{pmatrix}$

33. A factory for the construction of quality furniture has two divisions: a machine shop where the parts of the furniture are fabricated, and an assembly and finishing division where the parts are put together into the finished product. Suppose there are 12 employees in the machine shop and 20 in the assembly and finishing division and that each employee works an 8-hour day. Suppose further that the factory produces only two products: chairs and tables. A chair requires $\frac{384}{17}$ hours of machine shop time and $\frac{480}{17}$ hours of assembly and finishing time. A table requires $\frac{240}{17}$ hours of machine shop time and $\frac{640}{17}$ hours of assembly and finishing time. Assuming that there is an unlimited demand for these products and that the manufacturer wishes to keep all employees busy, how many chairs and how many tables can this factory produce each day?

34. A witch's magic cupboard contains 10 oz of ground four-leaf clovers and 14 oz of powdered mandrake root. The cupboard will replenish itself automatically provided she uses up exactly all her supplies. A batch of love potion requires $3\frac{1}{13}$ oz of ground four-leaf clovers and $2\frac{2}{13}$ oz of powdered mandrake root. One recipe of a well-known (to witches) cure for the common cold requires $5\frac{5}{13}$ oz of four-leaf colvers and $10\frac{10}{13}$ oz of mandrake root. How much of the love potion and the cold remedy should the witch make in order to use up the supply in the cupboard exactly?

35. A farmer feeds his cattle a mixture of two types of feed. One standard unit of type A feed supplies a steer with 10% of its minimum daily requirement of protein and 15% of its requirement of carbohydrates. Type B feed contains 12% of the requirement of protein and 8% of the requirement of carbohydrates in a standard unit. If the farmer wishes to feed his cattle exactly 100% of their minimum daily requirement of protein and carbohydrates, how many units of each type of feed should he give a steer each day?

36. A much simplified version of an input-output table for the 1958 Israeli economy divides that economy into three sectors—agriculture, manufacturing, and energy—with the following result.[†]

	Agriculture	Manufacturing	Energy
Agriculture	0.293	0	0
Manufacturing	0.014	0.207	0.017
Energy	0.044	0.010	0.216

(a) How many units of agricultural production are required to produce one unit of agricultural output?

(b) How many units of agricultural production are required to produce 200,000 units of agricultural output?

[†] Wassily Leontief, *Input-Output Economics* (New York: Oxford University Press, 1966), 54–57.

(c) How many units of agricultural product go into the production of 50,000 units of energy?

(d) How many units of energy go into the production of 50,000 units of agricultural products?

37. Continuing Problem 36, exports (in thousands of Israeli pounds) in 1958 were

Agriculture	13,213
Manufacturing	17,597
Energy	1,786

(a) Compute the technology and Leontief matrices.

(b) Determine the number of Israeli pounds worth of agricultural products, manufactured goods, and energy required to run this model of the Israeli economy and export the stated value of products.

38. Let $A = \begin{pmatrix} a & b \\ c & d \end{pmatrix}$. Show that $\det A \neq 0$ if and only if the rows of A are linearly independent. [*Hint:* See Theorem 2.6.1, page 54]

In Problems 39–45 compute the row echelon form of the given matrix and use it to determine directly whether the given matrix is invertible.

39. The matrix of Problem 1.
40. The matrix of Problem 4.
41. The matrix of Problem 7.
42. The matrix of Problem 9.
43. The matrix of Problem 11.
44. The matrix of Problem 13.
45. The matrix of Problem 14.

46. Let $A = \begin{pmatrix} a_{11} & a_{12} \\ a_{21} & a_{22} \end{pmatrix}$ and assume that $a_{11}a_{22} - a_{12}a_{21} \neq 0$. Derive formula (12) by row reducing the augmented matrix $\begin{pmatrix} a_{11} & a_{12} & | & 1 & 0 \\ a_{21} & a_{22} & | & 0 & 1 \end{pmatrix}$.

47. Prove parts (*i*) and (*ii*) of Theorem 5.

2.8 The Transpose of a Matrix

Corresponding to every matrix is another matrix which, as we shall see in Chapter 3, has properties very similar to those of the original matrix.

DEFINITION 1   **TRANSPOSE**   Let $A = (a_{ij})$ be an $m \times n$ matrix. Then the **transpose** of A, written A^t, is the $n \times m$ matrix obtained by interchanging the rows and columns of A. Succinctly, we may write $A^t = (a_{ji})$. In other words,

$$\text{if } A = \begin{pmatrix} a_{11} & a_{12} & \cdots & a_{1n} \\ a_{21} & a_{22} & \cdots & a_{2n} \\ \vdots & \vdots & & \vdots \\ a_{m1} & a_{m2} & \cdots & a_{mn} \end{pmatrix}, \text{ then } A^t = \begin{pmatrix} a_{11} & a_{21} & \cdots & a_{m1} \\ a_{12} & a_{22} & \cdots & a_{m2} \\ \vdots & \vdots & & \vdots \\ a_{1n} & a_{2n} & \cdots & a_{mn} \end{pmatrix} \tag{1}$$

Simply put, the ith row of A is the ith column of A^t and the jth column of A is the jth row of A^t.

EXAMPLE 1   Find the transposes of the matrices

$$A = \begin{pmatrix} 2 & 3 \\ 1 & 4 \end{pmatrix} \quad B = \begin{pmatrix} 2 & 3 & 1 \\ -1 & 4 & 6 \end{pmatrix} \quad C = \begin{pmatrix} 1 & 2 & -6 \\ 2 & -3 & 4 \\ 0 & 1 & 2 \\ 2 & -1 & 5 \end{pmatrix}$$

Solution   Interchanging the rows and columns of each matrix, we obtain

$$A^t = \begin{pmatrix} 2 & 1 \\ 3 & 4 \end{pmatrix} \quad B^t = \begin{pmatrix} 2 & -1 \\ 3 & 4 \\ 1 & 6 \end{pmatrix} \quad C^t = \begin{pmatrix} 1 & 2 & 0 & 2 \\ 2 & -3 & 1 & -1 \\ -6 & 4 & 2 & 5 \end{pmatrix}$$

Note, for example, that 4 is the component in row 2 and column 3 of C while 4 is the component in row 3 and column 2 of C^t. That is, the 23 element of C is the 32 element of C^t.

THEOREM 1   Suppose $A = (a_{ij})$ is an $n \times m$ matrix and $B = (b_{ij})$ is an $m \times p$ matrix. Then:

i. $(A^t)^t = A$. $\quad(2)$
ii. $(AB)^t = B^t A^t$ $\quad(3)$
iii. If A and B are $n \times m$, then $(A+B)^t = A^t + B^t$. $\quad(4)$

Proof   **i.** This follows directly from the definition of the transpose.
ii. First we note that AB is an $n \times p$ matrix, so $(AB)^t$ is $p \times n$. Also, B^t is $p \times m$ and A^t is $m \times n$, so $B^t A^t$ is $p \times n$. Thus both matrices in Equation (3) have the same size. Now the ijth element of AB is $\sum_{k=1}^{m} a_{ik} b_{kj}$ and this is the jith element of $(AB)^t$. Let $C = B^t$ and $D = A^t$. Then the ijth element c_{ij} of C is b_{ji} and the ijth element d_{ji} of D is a_{ji}. Thus the jith element of $CD = $ the jith element of $B^t A^t = \sum_{k=1}^{m} c_{jk} d_{ki} = \sum_{k=1}^{m} b_{kj} a_{ik} = \sum_{k=1}^{m} a_{ik} b_{kj} = $ the jith element of $(AB)^t$. This completes the proof of part (ii).
iii. This part is left as an exercise (see Problem 11). ∎

The transpose plays an important role in matrix theory. We shall see in succeeding chapters that A and A^t have many properties in common. Since columns of A^t are rows of A, we shall be able to use facts about the transpose to conclude that just about anything which is true about the rows of a matrix is true about its columns. We conclude this section with an important definition.

DEFINITION 2 **SYMMETRIC MATRIX** The $n \times n$ (square) matrix A is called **symmetric** if $A^t = A$.

EXAMPLE 2 The following four matrices are symmetric:

$$A = \begin{pmatrix} 1 & 2 \\ 2 & 3 \end{pmatrix} \quad B = \begin{pmatrix} 1 & -4 & 2 \\ -4 & 7 & 5 \\ 2 & 5 & 0 \end{pmatrix} \quad C = \begin{pmatrix} -1 & 2 & 4 & 6 \\ 2 & 7 & 3 & 5 \\ 4 & 3 & 8 & 0 \\ 6 & 5 & 0 & -4 \end{pmatrix}$$

We shall see the importance of symmetric matrices in Chapters 6 and 7.

PROBLEMS 2.8 In Problems 1–10 find the transpose of the given matrix.

1. $\begin{pmatrix} -1 & 4 \\ 6 & 5 \end{pmatrix}$
2. $\begin{pmatrix} 3 & 0 \\ 1 & 2 \end{pmatrix}$
3. $\begin{pmatrix} 2 & 3 \\ -1 & 2 \\ 1 & 4 \end{pmatrix}$
4. $\begin{pmatrix} 2 & -1 & 0 \\ 1 & 5 & 6 \end{pmatrix}$

5. $\begin{pmatrix} 1 & 2 & 3 \\ -1 & 0 & 4 \\ 1 & 5 & 5 \end{pmatrix}$
6. $\begin{pmatrix} 1 & 2 & 3 \\ 2 & 4 & -5 \\ 3 & -5 & 7 \end{pmatrix}$
7. $\begin{pmatrix} 1 & 0 & 1 & 0 \\ 0 & 1 & 0 & 1 \end{pmatrix}$
8. $\begin{pmatrix} 2 & -1 \\ 2 & 4 \\ 1 & 6 \\ 1 & 5 \end{pmatrix}$

9. $\begin{pmatrix} a & b & c \\ d & e & f \\ g & h & j \end{pmatrix}$
10. $\begin{pmatrix} 0 & 0 & 0 \\ 0 & 0 & 0 \end{pmatrix}$

11. Let A and B be $n \times m$ matrices. Show, using Definition 1, that $(A+B)^t = A^t + B^t$.

12. Find numbers α and β such that $\begin{pmatrix} 2 & \alpha & 3 \\ 5 & -6 & 2 \\ \beta & 2 & 4 \end{pmatrix}$ is symmetric.

13. If A and B are symmetric $n \times n$ matrices, prove that $A+B$ is symmetric.
14. If A and B are symmetric $n \times n$ matrices, show that $(AB)^t = BA$.
15. For any matrix A, show that the product matrix AA^t is defined and is a symmetric matrix.
16. Show that every diagonal matrix (see Problem 2.7.25, page 75) is symmetric.
17. Show that the transpose of every upper triangular matrix (see Problem 2.7.29) is lower triangular.

18. A square matrix is called **skew-symmetric** if $A^t = -A$ (that is, $a_{ij} = -a_{ji}$). Which of the following matrices are skew-symmetric?

a. $\begin{pmatrix} 1 & -6 \\ 6 & 0 \end{pmatrix}$
b. $\begin{pmatrix} 0 & -6 \\ 6 & 0 \end{pmatrix}$
c. $\begin{pmatrix} 2 & -2 & -2 \\ 2 & 2 & -2 \\ 2 & 2 & 2 \end{pmatrix}$
d. $\begin{pmatrix} 0 & 1 & -1 \\ -1 & 0 & 2 \\ 1 & -2 & 0 \end{pmatrix}$

19. Let A and B be $n \times n$ skew-symmetric matrices. Show that $A + B$ is skew-symmetric.

20. If A is skew-symmetric, show that every component on the main diagonal of A is zero.

21. If A and B are skew-symmetric $n \times n$ matrices, show that $(AB)^t = BA$, so that AB is symmetric if and only if A and B commute.

*22. Let $A = \begin{pmatrix} a_{11} & a_{12} \\ a_{21} & a_{22} \end{pmatrix}$ be a matrix with nonnegative entries having the properties that (i) $a_{11}^2 + a_{21}^2 = 1$ and $a_{12}^2 + a_{22}^2 = 1$ and (ii) $\begin{pmatrix} a_{11} \\ a_{21} \end{pmatrix} \cdot \begin{pmatrix} a_{12} \\ a_{22} \end{pmatrix} = 0$. Show that A is invertible and that $A^{-1} = A^t$.

Review Exercises for Chapter 2

In Exercises 1–7 compute using $\mathbf{a} = \begin{pmatrix} -1 \\ 4 \\ 6 \end{pmatrix}$, $\mathbf{b} = \begin{pmatrix} 5 \\ 2 \\ -1 \end{pmatrix}$, and $\mathbf{c} = \begin{pmatrix} -1 \\ 3 \\ 6 \end{pmatrix}$.

1. $\mathbf{a} + \mathbf{b}$
2. $2\mathbf{a} - 3\mathbf{c}$
3. $\mathbf{a} + \mathbf{b} - 2\mathbf{c}$
4. $-4\mathbf{a} + 3\mathbf{b} + 5\mathbf{c}$
5. $\mathbf{a} \cdot \mathbf{b}$
6. $\mathbf{b} \cdot \mathbf{c}$
7. $(\mathbf{a} - \mathbf{b}) \cdot (2\mathbf{c} - 3\mathbf{b})$

8. Find a number α such that $\begin{pmatrix} 2 \\ -1 \\ 4 \\ 6 \end{pmatrix}$ and $\begin{pmatrix} 1 \\ 5 \\ \alpha \\ 4 \end{pmatrix}$ are orthogonal.

In Exercises 9–16 perform the indicated computations.

9. $3 \begin{pmatrix} -2 & 1 \\ 0 & 4 \\ 2 & 3 \end{pmatrix}$

10. $\begin{pmatrix} 1 & 0 & 3 \\ 2 & -1 & 6 \end{pmatrix} + \begin{pmatrix} 2 & 0 & 4 \\ -2 & 5 & 8 \end{pmatrix}$

11. $5 \begin{pmatrix} 2 & 1 & 3 \\ -1 & 2 & 4 \\ -6 & 1 & 5 \end{pmatrix} - 3 \begin{pmatrix} -2 & 1 & 4 \\ 5 & 0 & 7 \\ 2 & -1 & 3 \end{pmatrix}$

12. $\begin{pmatrix} 2 & 3 \\ -1 & 4 \end{pmatrix} \begin{pmatrix} 5 & -1 \\ 2 & 7 \end{pmatrix}$

13. $\begin{pmatrix} 2 & 3 & 1 & 5 \\ 0 & 6 & 2 & 4 \end{pmatrix} \begin{pmatrix} 5 & 7 & 1 \\ 2 & 0 & 3 \\ 1 & 0 & 0 \\ 0 & 5 & 6 \end{pmatrix}$

14. $\begin{pmatrix} 2 & 3 & 5 \\ -1 & 6 & 4 \\ 1 & 0 & 6 \end{pmatrix} \begin{pmatrix} 0 & -1 & 2 \\ 3 & 1 & 2 \\ -7 & 3 & 5 \end{pmatrix}$

15. $\begin{pmatrix} 1 & 0 & 3 & -1 & 5 \\ 2 & 1 & 6 & 2 & 5 \end{pmatrix} \begin{pmatrix} 7 & 1 \\ 2 & 3 \\ -1 & 0 \\ 5 & 6 \\ 2 & 3 \end{pmatrix}$

16. $\begin{pmatrix} 1 & -1 & 2 \\ 3 & 5 & 6 \\ 2 & 4 & -1 \end{pmatrix} \begin{pmatrix} 2 \\ 1 \\ 3 \end{pmatrix}$

17. Verify the associative law of matrix multiplication for the matrices
$$A = \begin{pmatrix} 2 & 3 & 1 \\ 0 & 4 & 6 \end{pmatrix}, \quad B = \begin{pmatrix} 1 & 0 & 2 \\ 0 & 3 & 3 \\ 5 & 1 & -1 \end{pmatrix}, \text{ and } C = \begin{pmatrix} 5 & 6 \\ -1 & 2 \\ 0 & 1 \end{pmatrix}.$$

In Exercises 18–22 determine whether the given set of vectors is linearly dependent or independent.

18. $\begin{pmatrix} 2 \\ 3 \end{pmatrix}; \begin{pmatrix} 4 \\ -6 \end{pmatrix}$

19. $\begin{pmatrix} 2 \\ 3 \end{pmatrix}; \begin{pmatrix} 4 \\ 6 \end{pmatrix}$

20. $\begin{pmatrix} 1 \\ -1 \\ 2 \end{pmatrix}; \begin{pmatrix} 3 \\ 0 \\ 1 \end{pmatrix}; \begin{pmatrix} 0 \\ 0 \\ 0 \end{pmatrix}$

21. $\begin{pmatrix} 1 \\ -4 \\ 2 \end{pmatrix}; \begin{pmatrix} 0 \\ 2 \\ -1 \end{pmatrix}; \begin{pmatrix} 2 \\ -10 \\ 5 \end{pmatrix}$

22. $\begin{pmatrix} 1 \\ 0 \\ 0 \\ 0 \end{pmatrix}; \begin{pmatrix} 0 \\ 1 \\ 0 \\ 0 \end{pmatrix}; \begin{pmatrix} 0 \\ 0 \\ 1 \\ 0 \end{pmatrix}; \begin{pmatrix} 0 \\ 0 \\ 0 \\ 1 \end{pmatrix}$

In Exercises 23–27 calculate the row echelon form and the inverse of the given matrix (if the inverse exists).

23. $\begin{pmatrix} 2 & 3 \\ -1 & 4 \end{pmatrix}$

24. $\begin{pmatrix} -1 & 2 \\ 2 & -4 \end{pmatrix}$

25. $\begin{pmatrix} 1 & 2 & 0 \\ 2 & 1 & -1 \\ 3 & 1 & 1 \end{pmatrix}$

26. $\begin{pmatrix} -1 & 2 & 0 \\ 4 & 1 & -3 \\ 2 & 5 & -3 \end{pmatrix}$

27. $\begin{pmatrix} 2 & 0 & 4 \\ -1 & 3 & 1 \\ 0 & 1 & 2 \end{pmatrix}$

In Exercises 28–30 first write the system in the form $A\mathbf{x} = \mathbf{b}$, then calculate A^{-1}, and, finally, use matrix multiplication to obtain the solution vector.

28. $\begin{aligned} x_1 - 3x_2 &= 4 \\ 2x_1 + 5x_2 &= 7 \end{aligned}$

29. $\begin{aligned} x_1 + 2x_2 &= 3 \\ 2x_1 + x_2 - x_3 &= -1 \\ 3x_1 + x_2 + x_3 &= 7 \end{aligned}$

30. $\begin{aligned} 2x_1 + 4x_3 &= 7 \\ -x_1 + 3x_2 + x_3 &= -4 \\ x_2 + 2x_3 &= 5 \end{aligned}$

In Exercises 31–36 calculate the transpose of the given matrix and determine whether the matrix is symmetric or skew-symmetric.†

31. $\begin{pmatrix} 2 & 3 & 1 \\ -1 & 0 & 2 \end{pmatrix}$

32. $\begin{pmatrix} 4 & 6 \\ 6 & 4 \end{pmatrix}$

33. $\begin{pmatrix} 2 & 3 & 1 \\ 3 & -6 & -5 \\ 1 & -5 & 9 \end{pmatrix}$

34. $\begin{pmatrix} 0 & 5 & 6 \\ -5 & 0 & 4 \\ -6 & -4 & 0 \end{pmatrix}$

35. $\begin{pmatrix} 1 & -1 & 4 & 6 \\ -1 & 2 & 5 & 7 \\ 4 & 5 & 3 & -8 \\ 6 & 7 & -8 & 9 \end{pmatrix}$

36. $\begin{pmatrix} 0 & 1 & -1 & 1 \\ -1 & 0 & 1 & -2 \\ 1 & 1 & 0 & 1 \\ 1 & -2 & -1 & 0 \end{pmatrix}$

---

† From Problem 2.8.18 on page 80 we have: A is skew-symmetric if $A^t = -A$.

3 Determinants

3.1 Definitions

Let $A = \begin{pmatrix} a_{11} & a_{12} \\ a_{21} & a_{22} \end{pmatrix}$ be a 2×2 matrix. In Section 2.7 on page 65 we defined the determinant of A by

$$\det A = a_{11}a_{22} - a_{12}a_{21} \tag{1}$$

We shall often denote $\det A$ by

$$|A| = \begin{vmatrix} a_{11} & a_{12} \\ a_{21} & a_{22} \end{vmatrix} \tag{2}$$

We showed that A is invertible if and only if $\det A \neq 0$. As we shall see, this important theorem is valid for $n \times n$ matrices.

In this chapter we shall develop some of the basic properties of determinants and see how they can be used to calculate inverses and solve systems of n linear equations in n unknowns.

We shall define the determinant of an $n \times n$ matrix *inductively*. In other words, we use our knowledge of a 2×2 determinant to define a 3×3 determinant, use this to define a 4×4 determinant, and so on. We start by defining a 3×3 determinant.†

DEFINITION 1 **3×3 DETERMINANT** Let $A = \begin{pmatrix} a_{11} & a_{12} & a_{13} \\ a_{21} & a_{22} & a_{23} \\ a_{31} & a_{32} & a_{33} \end{pmatrix}$. Then

$$\det A = |A| = a_{11}\begin{vmatrix} a_{22} & a_{23} \\ a_{32} & a_{33} \end{vmatrix} - a_{12}\begin{vmatrix} a_{21} & a_{23} \\ a_{31} & a_{33} \end{vmatrix} + a_{13}\begin{vmatrix} a_{21} & a_{22} \\ a_{31} & a_{32} \end{vmatrix} \tag{3}$$

Note the minus sign before the second term on the right side of (3).

† There are several ways to define a determinant and this is one of them. It is important to realize that "det" is a function which assigns a *number* to a *square* matrix.

EXAMPLE 1   Let $A = \begin{pmatrix} 3 & 5 & 2 \\ 4 & 2 & 3 \\ -1 & 2 & 4 \end{pmatrix}$. Calculate $|A|$.

Solution

$$|A| = \begin{vmatrix} 3 & 5 & 2 \\ 4 & 2 & 3 \\ -1 & 2 & 4 \end{vmatrix} = 3\begin{vmatrix} 2 & 3 \\ 2 & 4 \end{vmatrix} - 5\begin{vmatrix} 4 & 3 \\ -1 & 4 \end{vmatrix} + 2\begin{vmatrix} 4 & 2 \\ -1 & 2 \end{vmatrix}$$

$$= 3 \cdot 2 - 5 \cdot 19 + 2 \cdot 10 = -69$$

EXAMPLE 2   Calculate $\begin{vmatrix} 2 & -3 & 5 \\ 1 & 0 & 4 \\ 3 & -3 & 9 \end{vmatrix}$.

Solution

$$\begin{vmatrix} 2 & -3 & 5 \\ 1 & 0 & 4 \\ 3 & -3 & 9 \end{vmatrix} = 2\begin{vmatrix} 0 & 4 \\ -3 & 9 \end{vmatrix} - (-3)\begin{vmatrix} 1 & 4 \\ 3 & 9 \end{vmatrix} + 5\begin{vmatrix} 1 & 0 \\ 3 & -3 \end{vmatrix}$$

$$= 2 \cdot 12 + 3(-3) + 5(-3) = 0$$

There is a simpler method for calculating 3×3 determinants. From Equation (3) we have

$$\begin{vmatrix} a_{11} & a_{12} & a_{13} \\ a_{21} & a_{22} & a_{23} \\ a_{31} & a_{32} & a_{33} \end{vmatrix} = a_{11}(a_{22}a_{33} - a_{23}a_{32}) - a_{12}(a_{21}a_{33} - a_{23}a_{31}) + a_{13}(a_{21}a_{32} - a_{22}a_{31})$$

or $|A| = a_{11}a_{22}a_{33} + a_{12}a_{23}a_{31} + a_{13}a_{21}a_{32} - a_{13}a_{22}a_{31} - a_{12}a_{21}a_{33} - a_{11}a_{32}a_{23}$ (4)

We write A and adjoin to it its first two columns

$$\begin{vmatrix} a_{11} & a_{12} & a_{13} \\ a_{21} & a_{22} & a_{23} \\ a_{31} & a_{32} & a_{33} \end{vmatrix} \begin{matrix} a_{11} & a_{12} \\ a_{21} & a_{22} \\ a_{31} & a_{32} \end{matrix}$$

We then calculate the six products, put minus signs before the products with arrows pointing upward, and add. This gives the sum in Equation (4).

EXAMPLE 3   Calculate $\begin{vmatrix} 3 & 5 & 2 \\ 4 & 2 & 3 \\ -1 & 2 & 4 \end{vmatrix}$ by using this new method.

Solution   Writing $\begin{vmatrix} 3 & 5 & 2 \\ 4 & 2 & 3 \\ -1 & 2 & 4 \end{vmatrix} \begin{matrix} 3 & 5 \\ 4 & 2 \\ -1 & 2 \end{matrix}$ and multiplying as indicated, we obtain

$$|A| = (3)(2)(4) + (5)(3)(-1) + (2)(4)(2) - (-1)(2)(2) - 2(3)(3) - (4)(4)(5)$$

$$= 24 - 15 + 16 + 4 - 18 - 80 = -69.$$

Warning. The method given above will *not* work for $n \times n$ determinants if $n \neq 3$. If you try something analogous for 4×4 or higher-order determinants, you will get the wrong answer.

Before defining $n \times n$ determinants, we first note that in Equation (3), $\begin{pmatrix} a_{22} & a_{23} \\ a_{32} & a_{33} \end{pmatrix}$ is the matrix obtained by deleting the first row and first column of A; $\begin{pmatrix} a_{21} & a_{23} \\ a_{31} & a_{33} \end{pmatrix}$ is the matrix obtained by deleting the first row and second column of A; and $\begin{pmatrix} a_{21} & a_{22} \\ a_{31} & a_{32} \end{pmatrix}$ is the matrix obtained by deleting the first row and third column of A. If we denote these three matrices by M_{11}, M_{12}, and M_{13}, respectively, and if $A_{11} = \det M_{11}$, $A_{12} = -\det M_{12}$, and $A_{13} = \det M_{13}$, then Equation (3) can be written

$$\det A = |A| = a_{11}A_{11} + a_{12}A_{12} + a_{13}A_{13} \tag{5}$$

DEFINITION 2   **MINOR**   Let A be an $n \times n$ matrix and let M_{ij} be the $(n-1) \times (n-1)$ matrix obtained from A by deleting the ith row and jth column of A. M_{ij} is called the **ij th minor** of A.

EXAMPLE 4   Let $A = \begin{pmatrix} 2 & -1 & 4 \\ 0 & 1 & 5 \\ 6 & 3 & -4 \end{pmatrix}$. Find M_{13} and M_{32}.

Solution   Deleting the first row and third column of A, we obtain $M_{13} = \begin{pmatrix} 0 & 1 \\ 6 & 3 \end{pmatrix}$. Similarly, by eliminating the third row and second column we obtain $M_{32} = \begin{pmatrix} 2 & 4 \\ 0 & 5 \end{pmatrix}$.

EXAMPLE 5   Let $A = \begin{pmatrix} 1 & -3 & 5 & 6 \\ 2 & 4 & 0 & 3 \\ 1 & 5 & 9 & -2 \\ 4 & 0 & 2 & 7 \end{pmatrix}$ Find M_{32} and M_{24}.

Solution   Deleting the third row and second column of A, we find that $M_{32} = \begin{pmatrix} 1 & 5 & 6 \\ 2 & 0 & 3 \\ 4 & 2 & 7 \end{pmatrix}$; similarly, $M_{24} = \begin{pmatrix} 1 & -3 & 5 \\ 1 & 5 & 9 \\ 4 & 0 & 2 \end{pmatrix}$.

DEFINITION 3   **COFACTOR** Let A be an $n \times n$ matrix. The **ijth cofactor** of A, denoted A_{ij}, is given by

$$A_{ij} = (-1)^{i+j}|M_{ij}| \qquad (6)$$

That is, the *ij*th cofactor of A is obtained by taking the determinant of the *ij*th minor and multiplying it by $(-1)^{i+j}$. Note that

$$(-1)^{i+j} = \begin{cases} 1 & \text{if } i+j \text{ is even} \\ -1 & \text{if } i+j \text{ is odd} \end{cases}$$

Remark. Definition 3 makes sense because we are going to define an $n \times n$ determinant with the assumption that we already know what an $(n-1) \times (n-1)$ determinant is.

EXAMPLE 6   In Example 5 we have

$$A_{32} = (-1)^{3+2}|M_{32}| = -\begin{vmatrix} 1 & 5 & 6 \\ 2 & 0 & 3 \\ 4 & 2 & 7 \end{vmatrix} = -8$$

$$A_{24} = (-1)^{2+4}\begin{vmatrix} 1 & -3 & 5 \\ 1 & 5 & 9 \\ 4 & 0 & 2 \end{vmatrix} = -192$$

We now consider the general $n \times n$ matrix. Here

$$A = \begin{pmatrix} a_{11} & a_{12} & \cdots & a_{1n} \\ a_{21} & a_{22} & \cdots & a_{2n} \\ \vdots & \vdots & & \vdots \\ a_{n1} & a_{n2} & \cdots & a_{nn} \end{pmatrix} \qquad (7)$$

DEFINITION 4   **$n \times n$ DETERMINANT** Let A be an $n \times n$ matrix. Then the determinant of A, written det A or $|A|$, is given by

$$\det A = |A| = a_{11}A_{11} + a_{12}A_{12} + a_{13}A_{13} + \cdots + a_{1n}A_{1n}$$
$$= \sum_{k=1}^{n} a_{1k}A_{1k} \qquad (8)$$

The expression on the right side of (8) is called an **expansion of cofactors**.

In Equation (8) we defined the determinant by expanding by cofactors using components of A in the first row. We shall see in the next section (Theorem 3.2.1) that we get the same answer if we expand by cofactors in any row or column.

EXAMPLE 7 Calculate det A, where

$$A = \begin{pmatrix} 1 & 3 & 5 & 2 \\ 0 & -1 & 3 & 4 \\ 2 & 1 & 9 & 6 \\ 3 & 2 & 4 & 8 \end{pmatrix}$$

Solution

$$\begin{vmatrix} 1 & 3 & 5 & 2 \\ 0 & -1 & 3 & 4 \\ 2 & 1 & 9 & 6 \\ 3 & 2 & 4 & 8 \end{vmatrix} = a_{11}A_{11} + a_{12}A_{12} + a_{13}A_{13} + a_{14}A_{14}$$

$$= 1\begin{vmatrix} -1 & 3 & 4 \\ 1 & 9 & 6 \\ 2 & 4 & 8 \end{vmatrix} - 3\begin{vmatrix} 0 & 3 & 4 \\ 2 & 9 & 6 \\ 3 & 4 & 8 \end{vmatrix} + 5\begin{vmatrix} 0 & -1 & 4 \\ 2 & 1 & 6 \\ 3 & 2 & 8 \end{vmatrix} - 2\begin{vmatrix} 0 & -1 & 3 \\ 2 & 1 & 9 \\ 3 & 2 & 4 \end{vmatrix}$$

$$= 1(-92) - 3(-70) + 5(2) - 2(-16) = 160$$

It is clear that calculating the determinant of an $n \times n$ matrix can be tedious. To calculate a 4×4 determinant, we must calculate four 3×3 determinants. To calculate a 5×5 determinant, we must calculate five 4×4 determinants—which is the same as calculating twenty 3×3 determinants. Fortunately, techniques exist for greatly simplifying these computations. Some of these methods are discussed in the next section. There are, however, some matrices whose determinants can easily be calculated.

DEFINITION 5 A square matrix is called **upper triangular** if all its components below the diagonal are zero. It is **lower triangular** if all its components above the diagonal are zero. A matrix is called **diagonal** if all its elements not on the diagonal are zero; that is, $A = (a_{ij})$ is upper triangular if $a_{ij} = 0$ for $i > j$, lower triangular if $a_{ij} = 0$ for $i < j$, and diagonal if $a_{ij} = 0$ for $i \neq j$. Note that a diagonal matrix is both upper and lower triangular.

EXAMPLE 8 The matrices $A = \begin{pmatrix} 2 & 1 & 7 \\ 0 & 2 & -5 \\ 0 & 0 & 1 \end{pmatrix}$ and $B = \begin{pmatrix} -2 & 3 & 0 & 1 \\ 0 & 0 & 2 & 4 \\ 0 & 0 & 1 & 3 \\ 0 & 0 & 0 & -2 \end{pmatrix}$ are upper triangular; $C = \begin{pmatrix} 5 & 0 & 0 \\ 2 & 3 & 0 \\ -1 & 2 & 4 \end{pmatrix}$ and $D = \begin{pmatrix} 0 & 0 \\ 1 & 0 \end{pmatrix}$ are lower triangular; I and $E = \begin{pmatrix} 2 & 0 & 0 \\ 0 & -7 & 0 \\ 0 & 0 & -4 \end{pmatrix}$ are diagonal.

EXAMPLE 9

Let

$$A = \begin{pmatrix} a_{11} & 0 & 0 & 0 \\ a_{21} & a_{22} & 0 & 0 \\ a_{31} & a_{32} & a_{33} & 0 \\ a_{41} & a_{42} & a_{43} & a_{44} \end{pmatrix}$$

be lower triangular. Compute det A.

Solution

$$\det A = a_{11}A_{11} + 0A_{12} + 0A_{13} + 0A_{14} = a_{11}A_{11}$$

$$= a_{11}\begin{vmatrix} a_{22} & 0 & 0 \\ a_{32} & a_{33} & 0 \\ a_{42} & a_{43} & a_{44} \end{vmatrix}$$

$$= a_{11}a_{22}\begin{vmatrix} a_{33} & 0 \\ a_{43} & a_{44} \end{vmatrix}$$

$$= a_{11}a_{22}a_{33}a_{44}$$

Example 9 can easily be generalized to prove the following.

THEOREM 1

Let $A = (a_{ij})$ be an upper† or lower triangular $n \times n$ matrix. Then

$$\det A = a_{11}a_{22}a_{33}\cdots a_{nn} \qquad (9)$$

That is: The determinant of a triangular matrix equals the product of its diagonal components.

EXAMPLE 10

The determinants of the six matrices in Example 8 are $|A| = 2 \cdot 2 \cdot 1 = 4$; $|B| = (-2)(0)(1)(-2) = 0$; $|C| = 5 \cdot 3 \cdot 4 = 60$; $|D| = 0$; $|I| = 1$; $|E| = (2)(-7)(-4) = 56$.

PROBLEMS 3.1

In Problems 1–10 calculate the determinant.

1. $\begin{vmatrix} 1 & 0 & 3 \\ 0 & 1 & 4 \\ 2 & 1 & 0 \end{vmatrix}$

2. $\begin{vmatrix} -1 & 1 & 0 \\ 2 & 1 & 4 \\ 1 & 5 & 6 \end{vmatrix}$

3. $\begin{vmatrix} 3 & -1 & 4 \\ 6 & 3 & 5 \\ 2 & -1 & 6 \end{vmatrix}$

4. $\begin{vmatrix} -1 & 0 & 6 \\ 0 & 2 & 4 \\ 1 & 2 & -3 \end{vmatrix}$

5. $\begin{vmatrix} -2 & 3 & 1 \\ 4 & 6 & 5 \\ 0 & 2 & 1 \end{vmatrix}$

6. $\begin{vmatrix} 5 & -2 & 1 \\ 6 & 0 & 3 \\ -2 & 1 & 4 \end{vmatrix}$

† The proof for the upper triangular case is more difficult at this stage, but it will be just the same once we know that det A can be evaluated by expanding in any column (Theorem 3.2.1).

7. $\begin{vmatrix} 2 & 0 & 3 & 1 \\ 0 & 1 & 4 & 2 \\ 0 & 0 & 1 & 5 \\ 1 & 2 & 3 & 0 \end{vmatrix}$ 8. $\begin{vmatrix} -3 & 0 & 0 & 0 \\ -4 & 7 & 0 & 0 \\ 5 & 8 & -1 & 0 \\ 2 & 3 & 0 & 6 \end{vmatrix}$ 9. $\begin{vmatrix} -2 & 0 & 0 & 7 \\ 1 & 2 & -1 & 4 \\ 3 & 0 & -1 & 5 \\ 4 & 2 & 3 & 0 \end{vmatrix}$

10. $\begin{vmatrix} 2 & 3 & -1 & 4 & 5 \\ 0 & 1 & 7 & 8 & 2 \\ 0 & 0 & 4 & -1 & 5 \\ 0 & 0 & 0 & -2 & 8 \\ 0 & 0 & 0 & 0 & 6 \end{vmatrix}$

11. Show that if A and B are diagonal $n \times n$ matrices, then $\det AB = \det A \det B$.

*12. Show that if A and B are lower triangular matrices, then $\det AB = \det A \det B$.

13. Show that, in general, it is not true that $\det(A+B) = \det A + \det B$.

14. Show that if A is triangular, then $\det A \neq 0$ if and only if all the diagonal components of A are nonzero.

15. Prove Theorem 1 for a lower triangular matrix.

*16. We say that the vectors $\begin{pmatrix} 1 \\ 0 \end{pmatrix}$ and $\begin{pmatrix} 0 \\ 1 \end{pmatrix}$ *generate the area* 1 in the plane since if we construct a square with three of its vertices at $(0,0)$, $(1,0)$, and $(0,1)$, we see that the area is 1. (See Figure 3.1a.) More generally, if $\begin{pmatrix} x_1 \\ y_1 \end{pmatrix}$ and $\begin{pmatrix} x_2 \\ y_2 \end{pmatrix}$ are two linearly independent 2-vectors, then they generate an area defined to be the area of the parallelogram with three of its four vertices at $(0,0)$, (x_1, y_1), and (x_2, y_2). (See Figure 3.1b.)

Figure 3.1

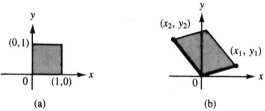

(a) (b)

Let A be a 2×2 matrix. If k denotes the area generated by $\begin{pmatrix} x_1 \\ y_1 \end{pmatrix}$ and $\begin{pmatrix} x_2 \\ y_2 \end{pmatrix}$, where $\begin{pmatrix} x_1 \\ y_1 \end{pmatrix} = A \begin{pmatrix} 1 \\ 0 \end{pmatrix}$ and $\begin{pmatrix} x_2 \\ y_2 \end{pmatrix} = A \begin{pmatrix} 0 \\ 1 \end{pmatrix}$, show that $k = |\det A|$.

**17. Let $\mathbf{u}_1$ and $\mathbf{u}_2$ be two 2-vectors and let $\mathbf{v}_1 = A\mathbf{u}_1$ and $\mathbf{v}_2 = A\mathbf{u}_2$. Show that

(area generated by $\mathbf{v}_1$ and $\mathbf{v}_2$) = (area generated by $\mathbf{u}_1$ and $\mathbf{u}_2$) $|\det A|$

This provides a geometric interpretation of the determinant.

3.2 Properties of Determinants

Determinants have many properties that can make computations easier. We begin to describe these properties by stating a theorem from which everything else follows. The proof of this theorem is difficult and is deferred to the next section.

THEOREM 1 **Basic Theorem.** Let

$$A = \begin{pmatrix} a_{11} & a_{12} & \cdots & a_{1n} \\ a_{21} & a_{22} & \cdots & a_{2n} \\ \vdots & \vdots & & \vdots \\ a_{n1} & a_{n2} & \cdots & a_{nn} \end{pmatrix}$$

be an $n \times n$ matrix. Then

$$\det A = a_{i1}A_{i1} + a_{i2}A_{i2} + \cdots + a_{in}A_{in} = \sum_{k=1}^{n} a_{ik}A_{ik} \qquad (1)$$

for $i = 1, 2, \ldots, n$. That is, we can calculate det A by expanding by cofactors in *any* row of A. Furthermore:

$$\det A = a_{1j}A_{1j} + a_{2j}A_{2j} + \cdots + a_{nj}A_{nj} = \sum_{k=1}^{n} a_{kj}A_{kj} \qquad (2)$$

Since the jth column of A is $\begin{pmatrix} a_{1j} \\ a_{2j} \\ \vdots \\ a_{nj} \end{pmatrix}$, Equation (2) indicates that we can

calculate det A by expanding by cofactors in any column of A.

EXAMPLE 1 For $A = \begin{pmatrix} 3 & 5 & 2 \\ 4 & 2 & 3 \\ -1 & 2 & 4 \end{pmatrix}$, we saw in Example 3.1.1 that det $A = -69$. Expanding in the second row we obtain

$$\det A = 4A_{21} + 2A_{22} + 3A_{23}$$

$$= 4(-1)^{2+1} \begin{vmatrix} 5 & 2 \\ 2 & 4 \end{vmatrix} + 2(-1)^{2+2} \begin{vmatrix} 3 & 2 \\ -1 & 4 \end{vmatrix} + 3(-1)^{2+3} \begin{vmatrix} 3 & 5 \\ -1 & 2 \end{vmatrix}$$

$$= -4(16) + 2(14) - 3(11) = -69$$

Similarly, if we expand in the third column, say, we obtain

$$\det A = 2A_{13} + 3A_{23} + 4A_{33}$$

$$= 2(-1)^{1+3} \begin{vmatrix} 4 & 2 \\ -1 & 2 \end{vmatrix} + 3(-1)^{2+3} \begin{vmatrix} 3 & 5 \\ -1 & 2 \end{vmatrix} + 4(-1)^{3+3} \begin{vmatrix} 3 & 5 \\ 4 & 2 \end{vmatrix}$$

$$= 2(10) - 3(11) + 4(-14) = -69$$

You should verify that we get the same answer if we expand in the third row or the first or second column.

We now list and prove some additional properties of determinants. In each case we assume that A is an $n \times n$ matrix.† We shall see that these properties can be used to reduce greatly the work involved in evaluating a determinant.

Property 1 If any row or column of A is the zero vector, then $\det A = 0$.

Proof Suppose the ith row of A contains all zeros. That is, $a_{ij} = 0$ for $j = 1, 2, \ldots, n$. Then $\det A = a_{i1}A_{i1} + a_{i2}A_{i2} + \cdots + a_{in}A_{in} = 0 + 0 + \cdots + 0 = 0$. The same proof works if the jth column is the zero vector. ∎

EXAMPLE 2 It is easy to verify that

$$\begin{vmatrix} 2 & 3 & 5 \\ 0 & 0 & 0 \\ 1 & -2 & 4 \end{vmatrix} = 0 \quad \text{and} \quad \begin{vmatrix} -1 & 3 & 0 & 1 \\ 4 & 2 & 0 & 5 \\ -1 & 6 & 0 & 4 \\ 2 & 1 & 0 & 1 \end{vmatrix} = 0$$

Property 2 If the ith row or the jth column of A is multiplied by the constant c, then $\det A$ is multiplied by c. That is, if we call this new matrix B, then

$$|B| = \begin{vmatrix} a_{11} & a_{12} & \cdots & a_{1n} \\ a_{21} & a_{22} & \cdots & a_{2n} \\ \vdots & \vdots & & \vdots \\ ca_{i1} & ca_{i2} & \cdots & ca_{in} \\ \vdots & \vdots & & \vdots \\ a_{n1} & a_{n2} & \cdots & a_{nn} \end{vmatrix} = c \begin{vmatrix} a_{11} & a_{12} & \cdots & a_{1n} \\ a_{21} & a_{22} & \cdots & a_{2n} \\ \vdots & \vdots & & \vdots \\ a_{i1} & a_{i2} & \cdots & a_{in} \\ \vdots & \vdots & & \vdots \\ a_{n1} & a_{n2} & \cdots & a_{nn} \end{vmatrix} = c|A| \quad (3)$$

Proof To prove (3) we expand in the ith row of A to obtain

$$\det B = ca_{i1}A_{i1} + ca_{i2}A_{i2} + \cdots ca_{in}A_{in}$$
$$= c(a_{i1}A_{i1} + a_{i2}A_{i2} + \cdots + a_{in}A_{in}) = c \det A$$

A similar proof works for columns. ∎

---

† The proofs of these properties are given in terms of the rows of a matrix. Using Theorem 1 the same properties can be proved for columns.

EXAMPLE 3 Let $A = \begin{pmatrix} 1 & -1 & 2 \\ 3 & 1 & 4 \\ 0 & -2 & 5 \end{pmatrix}$. Then $\det A = 16$. If we multiply the second row by 4, we have $B = \begin{pmatrix} 1 & -1 & 2 \\ 12 & 4 & 16 \\ 0 & -2 & 5 \end{pmatrix}$ and $\det B = 64 = 4 \det A$. If the third column is multiplied by -3, we obtain $C = \begin{pmatrix} 1 & -1 & -6 \\ 3 & 1 & -12 \\ 0 & -2 & -15 \end{pmatrix}$ and $\det C = -48 = -3 \det A$.

Remark. Using Property 2 we can prove (see Problem 28) the following interesting fact: For any scalar α and $n \times n$ matrix A, $\det \alpha A = \alpha^n \det A$.

Property 3 Let

$$A = \begin{pmatrix} a_{11} & a_{12} & \cdots & a_{1j} & \cdots & a_{1n} \\ a_{21} & a_{22} & \cdots & a_{2j} & \cdots & a_{2n} \\ \vdots & \vdots & & \vdots & & \vdots \\ a_{n1} & a_{n2} & \cdots & a_{nj} & \cdots & a_{nn} \end{pmatrix}, \quad B = \begin{pmatrix} a_{11} & a_{12} & \cdots & \alpha_{1j} & \cdots & a_{1n} \\ a_{21} & a_{22} & \cdots & \alpha_{2j} & \cdots & a_{2n} \\ \vdots & \vdots & & \vdots & & \vdots \\ a_{n1} & a_{n2} & \cdots & \alpha_{nj} & \cdots & a_{nn} \end{pmatrix},$$

and

$$C = \begin{pmatrix} a_{11} & a_{12} & \cdots & a_{1j}+\alpha_{1j} & \cdots & a_{1n} \\ a_{21} & a_{22} & \cdots & a_{2j}+\alpha_{2j} & \cdots & a_{2n} \\ \vdots & \vdots & & \vdots & & \vdots \\ a_{n1} & a_{n2} & \cdots & a_{nj}+\alpha_{nj} & \cdots & a_{nn} \end{pmatrix}$$

Then

$$\det C = \det A + \det B \tag{4}$$

In other words, suppose that A, B, and C are identical except for the jth column and that the jth column of C is the sum of the jth columns of A and B. Then $\det C = \det A + \det B$. The same statement is true for rows.

Proof We expand $\det C$ in the jth column to obtain

$$\det C = (a_{1j}+\alpha_{1j})A_{1j} + (a_{2j}+\alpha_{2j})A_{2j} + \cdots + (a_{nj}+\alpha_{nj})A_{nj}$$
$$= (a_{1j}A_{1j} + a_{2j}A_{2j} + \cdots + a_{nj}A_{nj})$$
$$+ (\alpha_{1j}A_{1j} + \alpha_{2j}A_{2j} + \cdots + \alpha_{nj}A_{nj}) = \det A + \det B \quad \blacksquare$$

EXAMPLE 4 Let $A = \begin{pmatrix} 1 & -1 & 2 \\ 3 & 1 & 4 \\ 0 & -2 & 5 \end{pmatrix}$, $B = \begin{pmatrix} 1 & -6 & 2 \\ 3 & 2 & 4 \\ 0 & 4 & 5 \end{pmatrix}$, and $C = \begin{pmatrix} 1 & -1-6 & 2 \\ 3 & 1+2 & 4 \\ 0 & -2+4 & 5 \end{pmatrix} = $

$\begin{pmatrix} 1 & -7 & 2 \\ 3 & 3 & 4 \\ 0 & 2 & 5 \end{pmatrix}$. Then det $A = 16$, det $B = 108$, and det $C = 124 = $ det $A +$ det B.

Property 4 Interchanging any two rows (or columns) of A has the effect of multiplying det A by -1.

Proof We prove the statement for rows and assume first that two adjacent rows are interchanged. That is, we assume that the ith and $(i+1)$st rows are interchanged. Let

$$A = \begin{pmatrix} a_{11} & a_{12} & \cdots & a_{1n} \\ a_{21} & a_{22} & \cdots & a_{2n} \\ \vdots & \vdots & & \vdots \\ a_{i1} & a_{i2} & \cdots & a_{in} \\ a_{i+1,1} & a_{i+1,2} & \cdots & a_{i+1,n} \\ \vdots & \vdots & & \vdots \\ a_{n1} & a_{n2} & \cdots & a_{nn} \end{pmatrix} \quad \text{and} \quad B = \begin{pmatrix} a_{11} & a_{12} & \cdots & a_{1n} \\ a_{21} & a_{22} & \cdots & a_{2n} \\ \vdots & \vdots & & \vdots \\ a_{i+1,1} & a_{i+1,2} & \cdots & a_{i+1,n} \\ a_{i1} & a_{i2} & \cdots & a_{in} \\ \vdots & \vdots & & \vdots \\ a_{n1} & a_{n2} & \cdots & a_{nn} \end{pmatrix}$$

Then, expanding det A in its ith row and det B in its $(i+1)$st row, we obtain

$$\begin{aligned} \det A &= a_{i1}A_{i1} + a_{i2}A_{i2} + \cdots + a_{in}A_{in} \\ \det B &= a_{i1}B_{i+1,1} + a_{i2}B_{i+1,2} + \cdots + a_{in}B_{i+1,n} \end{aligned} \qquad (5)$$

Here $A_{ij} = (-1)^{i+j}|M_{ij}|$, where M_{ij} is obtained by crossing off the ith row and jth column of A. Notice now that if we cross off the $(i+1)$st row and jth column of B, we obtain the same M_{ij}. Thus

$$B_{i+1,j} = (-1)^{i+1+j}|M_{ij}| = -(-1)^{i+j}|M_{ij}| = -A_{ij}$$

so that, from Equations (5), det $B = -$det A.

Now suppose that $i < j$ and that the ith and jth rows are to be interchanged. We can do this by interchanging adjacent rows several times. It will take $j - i$ interchanges to move row j into the ith row. Then row i will be in the $(i+1)$st row and it will take an additional $j - i - 1$ interchanges to move row i into the jth row. To illustrate, we interchange rows 2 and 6:†

$$\begin{matrix}1\\2\\3\\4\\5\\6\\7\end{matrix} \to \begin{matrix}1\\2\\3\\4\\6\\5\\7\end{matrix} \to \begin{matrix}1\\2\\3\\6\\4\\5\\7\end{matrix} \to \begin{matrix}1\\2\\6\\3\\4\\5\\7\end{matrix} \to \begin{matrix}1\\6\\2\\3\\4\\5\\7\end{matrix} \to \begin{matrix}1\\6\\3\\2\\4\\5\\7\end{matrix} \to \begin{matrix}1\\6\\3\\4\\2\\5\\7\end{matrix} \to \begin{matrix}1\\6\\3\\4\\5\\2\\7\end{matrix}$$

$\underbrace{\hphantom{xxxxxxxxxxxxxxxxxxxxxxxxx}}_{\substack{6-2=4 \text{ interchanges to} \\ \text{move the 6 into the 2 position}}}$ $\underbrace{\hphantom{xxxxxxxxxxxxxxxxxxxx}}_{\substack{6-2-1=3 \text{ interchanges to get} \\ \text{the 2 into the 6 position}}}$

Finally, the total number of interchanges of adjacent rows is $(j-i)+(j-i-1)=2j-2i-1$, which is odd. Thus det A is multiplied by -1 an odd number of times, which is what we needed to show. ∎

EXAMPLE 5 Let $A = \begin{pmatrix} 1 & -1 & 2 \\ 3 & 1 & 4 \\ 0 & -2 & 5 \end{pmatrix}$. By interchanging the first and third rows we obtain $B = \begin{pmatrix} 0 & -2 & 5 \\ 3 & 1 & 4 \\ 1 & -1 & 2 \end{pmatrix}$. By interchanging the first and second columns of A we obtain $C = \begin{pmatrix} -1 & 1 & 2 \\ 1 & 3 & 4 \\ -2 & 0 & 5 \end{pmatrix}$. Then, by direct calculation, we find that det $A = 16$ and det $B = $ det $C = -16$.

Property 5 If A has two equal rows or columns, then det $A = 0$.

Proof Suppose the ith and jth rows of A are equal. By interchanging these rows we get a matrix B having the property that det $B = -$det A (from Property 4). But since row $i = $ row j, interchanging them gives us the same matrix. Thus $A = B$ and det $A = $ det $B = -$det A. Thus 2 det $A = 0$, which can happen only if det $A = 0$. ∎

EXAMPLE 6 By direct calculation we can verify that for $A = \begin{pmatrix} 1 & -1 & 2 \\ 5 & 7 & 3 \\ 1 & -1 & 2 \end{pmatrix}$ [two equal rows] and $B = \begin{pmatrix} 5 & 2 & 2 \\ 3 & -1 & -1 \\ -2 & 4 & 4 \end{pmatrix}$ [two equal columns], det $A = $ det $B = 0$.

† Note that all the numbers here refer to rows.

Property 6 If one row (column) of A is a constant multiple of another row (column), then $\det A = 0$.

Proof Let $(a_{j1}, a_{j2}, \ldots, a_{jn}) = c(a_{i1}, a_{i2}, \ldots, a_{in})$. Then, from Property 2,

$$\det A = c \begin{vmatrix} a_{11} & a_{12} & \cdots & a_{1n} \\ a_{21} & a_{22} & \cdots & a_{2n} \\ \vdots & \vdots & & \vdots \\ a_{i1} & a_{i2} & \cdots & a_{in} \\ \vdots & \vdots & & \vdots \\ a_{i1} & a_{i2} & \cdots & a_{in} \\ \vdots & \vdots & & \vdots \\ a_{n1} & a_{n2} & \cdots & a_{nn} \end{vmatrix} = 0 \quad \text{(from Property 5)}$$

where the jth row $\to a_{i1}\ a_{i2}\ \cdots\ a_{in}$. ∎

EXAMPLE 7

$\begin{vmatrix} 2 & -3 & 5 \\ 1 & 7 & 2 \\ -4 & 6 & -10 \end{vmatrix} = 0$ since the third row is -2 times the first row.

EXAMPLE 8

$\begin{vmatrix} 2 & 4 & 1 & 12 \\ -1 & 1 & 0 & 3 \\ 0 & -1 & 9 & -3 \\ 7 & 3 & 6 & 9 \end{vmatrix} = 0$ since the fourth column is three times the second column.

Property 7 If a multiple of one row (column) of A is added to another row (column) of A, then the determinant is unchanged.

Proof Let B be the matrix obtained by adding c times the ith row of A to the jth row of A. Then

$$\det B = \begin{vmatrix} a_{11} & a_{12} & \cdots & a_{1n} \\ a_{21} & a_{22} & \cdots & a_{2n} \\ \vdots & \vdots & & \vdots \\ a_{i1} & a_{i2} & \cdots & a_{in} \\ \vdots & \vdots & & \vdots \\ a_{j1}+ca_{i1} & a_{j2}+ca_{i2} & \cdots & a_{jn}+ca_{in} \\ \vdots & \vdots & & \vdots \\ a_{n1} & a_{n2} & \cdots & a_{nn} \end{vmatrix}$$

(from Property 3) =
$$\begin{vmatrix} a_{11} & a_{12} & \cdots & a_{1n} \\ a_{21} & a_{22} & \cdots & a_{2n} \\ \cdot & \cdot & & \cdot \\ \cdot & \cdot & & \cdot \\ \cdot & \cdot & & \cdot \\ a_{i1} & a_{i2} & \cdots & a_{in} \\ \cdot & \cdot & & \cdot \\ \cdot & \cdot & & \cdot \\ \cdot & \cdot & & \cdot \\ a_{j1} & a_{j2} & \cdots & a_{jn} \\ \cdot & \cdot & & \cdot \\ \cdot & \cdot & & \cdot \\ \cdot & \cdot & & \cdot \\ a_{n1} & a_{n2} & \cdots & a_{nn} \end{vmatrix} + \begin{vmatrix} a_{11} & a_{12} & \cdots & a_{1n} \\ a_{21} & a_{22} & \cdots & a_{2n} \\ \cdot & \cdot & & \cdot \\ \cdot & \cdot & & \cdot \\ \cdot & \cdot & & \cdot \\ a_{i1} & a_{i2} & \cdots & a_{in} \\ \cdot & \cdot & & \cdot \\ \cdot & \cdot & & \cdot \\ \cdot & \cdot & & \cdot \\ ca_{i1} & ca_{i2} & \cdots & ca_{in} \\ \cdot & \cdot & & \cdot \\ \cdot & \cdot & & \cdot \\ \cdot & \cdot & & \cdot \\ a_{n1} & a_{n2} & \cdots & a_{nn} \end{vmatrix}$$

$= \det A + 0 = \det A$ (the zero comes from Property 6) ∎

EXAMPLE 9   Let $A = \begin{pmatrix} 1 & -1 & 2 \\ 3 & 1 & 4 \\ 0 & -2 & 5 \end{pmatrix}$. Then $\det A = 16$. If we multiply the third row by 4 and add it to the second row, we obtain a new matrix B given by

$$B = \begin{pmatrix} 1 & -1 & 2 \\ 3+4(0) & 1+4(-2) & 4+5(4) \\ 0 & -2 & 5 \end{pmatrix} = \begin{pmatrix} 1 & -1 & 2 \\ 3 & -7 & 24 \\ 0 & -2 & 5 \end{pmatrix}$$

and $\det B = 16 = \det A$.

The properties discussed above make it much easier to evaluate high-order determinants. We simply "row-reduce" the determinant, using Property 7, until the determinant is in an easily evaluated form. The most common goal will be to use Property 7 repeatedly until either (i) the new determinant has a row (column) of zeros or one row (column) a multiple of another row (column)—in which case the determinant is zero—or (ii) the new matrix is triangular so that its determinant is the product of its diagonal elements.

EXAMPLE 10   Calculate

$$|A| = \begin{vmatrix} 1 & 3 & 5 & 2 \\ 0 & -1 & 3 & 4 \\ 2 & 1 & 9 & 6 \\ 3 & 2 & 4 & 8 \end{vmatrix}$$

Solution   (See Example 3.1.7, page 86.)

There is already a zero in the first column, so it is simplest to reduce other elements in the first column to zero. We then continue to reduce, aiming for a triangular matrix:

Multiply the first row by -2 and add it to the third row and multiply the first row by -3 and add it to the fourth row.

$$|A| = \begin{vmatrix} 1 & 3 & 5 & 2 \\ 0 & -1 & 3 & 4 \\ 0 & -5 & -1 & 2 \\ 0 & -7 & -11 & 2 \end{vmatrix}$$

Multiply the second row by -5 and -7 and add it to the third and fourth rows, respectively.

$$= \begin{vmatrix} 1 & 3 & 5 & 2 \\ 0 & -1 & 3 & 4 \\ 0 & 0 & -16 & -18 \\ 0 & 0 & -32 & -26 \end{vmatrix}$$

Factor out -16 from the third row (using Property 2).

$$= -16 \begin{vmatrix} 1 & 3 & 5 & 2 \\ 0 & -1 & 3 & 4 \\ 0 & 0 & 1 & \frac{9}{8} \\ 0 & 0 & -32 & -26 \end{vmatrix}$$

Multiply the third row by 32 and add it to the fourth row.

$$= -16 \begin{vmatrix} 1 & 3 & 5 & 2 \\ 0 & -1 & 3 & 4 \\ 0 & 0 & 1 & \frac{9}{8} \\ 0 & 0 & 0 & 10 \end{vmatrix}$$

Now we have an upper triangular matrix and $|A| = -16(1)(-1)(1)(10) = (-16)(-10) = 160$.

EXAMPLE 11 Calculate

$$|A| = \begin{vmatrix} -2 & 1 & 0 & 4 \\ 3 & -1 & 5 & 2 \\ -2 & 7 & 3 & 1 \\ 3 & -7 & 2 & 5 \end{vmatrix}$$

Solution There are a number of ways to proceed here and it is not apparent which way will get us the answer most quickly. However, since there is already one zero in the first row, we begin our reduction in that row.

Multiply the second column by 2 and -4 and add it to the first and fourth columns, respectively.

$$|A| = \begin{vmatrix} 0 & 1 & 0 & 0 \\ 1 & -1 & 5 & 6 \\ 12 & 7 & 3 & -27 \\ -11 & -7 & 2 & 33 \end{vmatrix}$$

Interchange the first two columns.

$$= - \begin{vmatrix} 1 & 0 & 0 & 0 \\ -1 & 1 & 5 & 6 \\ 7 & 12 & 3 & -27 \\ -7 & -11 & 2 & 33 \end{vmatrix}$$

Multiply the second column by -5 and -6 and add it to the third and fourth columns, respectively.

$$= -\begin{vmatrix} 1 & 0 & 0 & 0 \\ -1 & 1 & 0 & 0 \\ 7 & 12 & -57 & -99 \\ -7 & -11 & 57 & 99 \end{vmatrix}$$

Since the fourth column is now a multiple of the third column (column $4 = \frac{99}{57} \times$ column 3), we see that $|A| = 0$.

EXAMPLE 12 Calculate

$$|A| = \begin{vmatrix} 1 & -2 & 3 & -5 & 7 \\ 2 & 0 & -1 & -5 & 6 \\ 4 & 7 & 3 & -9 & 4 \\ 3 & 1 & -2 & -2 & 3 \\ -5 & -1 & 3 & 7 & -9 \end{vmatrix}$$

Solution Adding first row 2 and then row 4 to row 5, we obtain

$$|A| = \begin{vmatrix} 1 & -2 & 3 & 5 & 7 \\ 2 & 0 & -1 & -5 & 6 \\ 4 & 7 & 3 & -9 & 4 \\ 3 & 1 & -2 & -2 & 3 \\ 0 & 0 & 0 & 0 & 0 \end{vmatrix} = 0 \quad \text{(from Property 1)}$$

This example illustrates the fact that a little looking before beginning the computations can simplify matters considerably.

There are three additional facts about determinants that will be very useful to us.

THEOREM 2 Let A be an $n \times n$ matrix. Then

$$a_{i1}A_{j1} + a_{i2}A_{j2} + \cdots + a_{in}A_{jn} = 0 \quad \text{if } i \neq j \tag{6}$$

Note. From Theorem 1 the sum in Equation (6) equals det A if $i = j$.

Proof Let

$$B = \begin{pmatrix} a_{11} & a_{12} & \cdots & a_{1n} \\ a_{21} & a_{22} & \cdots & a_{2n} \\ \vdots & \vdots & & \vdots \\ a_{i1} & a_{i2} & \cdots & a_{in} \\ \vdots & \vdots & & \vdots \\ a_{i1} & a_{i2} & \cdots & a_{in} \\ \vdots & \vdots & & \vdots \\ a_{n1} & a_{n2} & \cdots & a_{nn} \end{pmatrix} \quad j\text{th row} \rightarrow$$

Then, since two rows of B are equal, $\det B = 0$. But $B = A$ except in the jth row. Thus if we calculate $\det B$ by expanding in the jth row of B, we obtain the sum in (6) and the theorem is proved. Note that when we expand in the jth row, the jth row is deleted in computing the cofactors of B. Thus $B_{jk} = A_{ik}$ for $k = 1, 2, \ldots, n$. ∎

THEOREM 3 Let A be an $n \times n$ matrix. Then

$$\det A = \det A^t \qquad (7)$$

Proof This proof uses mathematical induction. If you are unfamiliar with this important method of proof, refer to Appendix 1. We first prove the theorem in the case $n = 2$. If

$$|A| = \begin{vmatrix} a_{11} & a_{12} \\ a_{21} & a_{22} \end{vmatrix} = a_{11}a_{22} - a_{12}a_{21}$$

then

$$|A^t| = \begin{vmatrix} a_{11} & a_{21} \\ a_{12} & a_{22} \end{vmatrix} = a_{11}a_{22} - a_{21}a_{12} = |A|$$

so the theorem is true for $n = 2$. Next we assume the theorem to be true for $(n-1) \times (n-1)$ matrices and prove it for $n \times n$ matrices. This will prove the theorem. Let $B = A^t$. Then

$$|A| = \begin{vmatrix} a_{11} & a_{12} & \cdots & a_{1n} \\ a_{21} & a_{22} & \cdots & a_{2n} \\ \vdots & \vdots & & \vdots \\ a_{n1} & a_{n2} & \cdots & a_{nn} \end{vmatrix} \quad \text{and} \quad |A^t| = |B| = \begin{vmatrix} a_{11} & a_{21} & \cdots & a_{n1} \\ a_{12} & a_{22} & \cdots & a_{n2} \\ \vdots & \vdots & & \vdots \\ a_{1n} & a_{2n} & \cdots & a_{nn} \end{vmatrix}$$

We expand $|A|$ in the first row and expand $|B|$ in the first column. This gives us

$$|A| = a_{11}A_{11} + a_{12}A_{12} + \cdots + a_{1n}A_{1n}$$
$$|B| = a_{11}B_{11} + a_{12}B_{21} + \cdots + a_{1n}B_{n1}$$

We need to show that $A_{1k} = B_{k1}$ for $k = 1, 2, \ldots, n$. But $A_{1k} = (-1)^{1+k}|M_{1k}|$ and $B_{k1} = (-1)^{k+1}|N_{k1}|$, where M_{1k} is the $1k$th minor of A and N_{k1} is the $k1$st minor of B. Then

$$|M_{1k}| = \begin{vmatrix} a_{21} & a_{22} & \cdots & a_{2,k-1} & a_{2,k+1} & \cdots & a_{2n} \\ a_{31} & a_{32} & \cdots & a_{3,k-1} & a_{3,k+1} & \cdots & a_{3n} \\ \vdots & \vdots & & \vdots & \vdots & & \vdots \\ a_{n1} & a_{n2} & \cdots & a_{n,k-1} & a_{n,k+1} & \cdots & a_{nn} \end{vmatrix}$$

and

$$|N_{k1}| = \begin{vmatrix} a_{21} & a_{31} & \cdots & a_{n1} \\ a_{22} & a_{32} & \cdots & a_{n2} \\ \vdots & \vdots & & \vdots \\ a_{2,k-1} & a_{3,k-1} & \cdots & a_{n,k-1} \\ a_{2,k+1} & a_{3,k+1} & \cdots & a_{n,k+1} \\ \vdots & \vdots & & \vdots \\ a_{2n} & a_{3n} & \cdots & a_{nn} \end{vmatrix}$$

Clearly $M_{1k} = N_{k1}^t$, and since both are $(n-1) \times (n-1)$ matrices, the induction hypothesis tells us that $|M_{1k}| = |N_{k1}|$. Thus $A_{1k} = B_{k1}$ and the proof is complete. ■

EXAMPLE 13 Let $A = \begin{pmatrix} 1 & -1 & 2 \\ 3 & 1 & 4 \\ 0 & -2 & 5 \end{pmatrix}$. Then $A^t = \begin{pmatrix} 1 & 3 & 0 \\ -1 & 1 & -2 \\ 2 & 4 & 5 \end{pmatrix}$ and it is easy to verify that $|A| = |A^t| = 16$.

THEOREM 4 Let A and B be $n \times n$ matrices. Then

$$\det AB = \det A \det B \qquad (8)$$

That is: *The determinant of the product is the product of the determinants.*

Remark. The proof of this theorem is not conceptually difficult, but, as you might imagine from having worked with matrix products, it is extremely

cumbersome. For that reason we shall simply prove the theorem in the case that A and B are 2×2 matrices. Let $A = \begin{pmatrix} a_{11} & a_{12} \\ a_{21} & a_{22} \end{pmatrix}$ and $B = \begin{pmatrix} b_{11} & b_{12} \\ b_{21} & b_{22} \end{pmatrix}$.
Then

$$\det A \det B = (a_{11}a_{22} - a_{12}a_{21})(b_{11}b_{22} - b_{12}b_{21})$$
$$= a_{11}a_{22}b_{11}b_{22} - a_{11}a_{22}b_{12}b_{21} - a_{12}a_{21}b_{11}b_{22} + a_{12}a_{21}b_{12}b_{21}$$

and $AB = \begin{pmatrix} a_{11}b_{11} + a_{12}b_{21} & a_{11}b_{12} + a_{12}b_{22} \\ a_{21}b_{11} + a_{22}b_{21} & a_{21}b_{12} + a_{22}b_{22} \end{pmatrix}$. Hence

$$\det AB = (a_{11}b_{11} + a_{12}b_{21})(a_{21}b_{12} + a_{22}b_{22}) - (a_{11}b_{12} + a_{12}b_{22})(a_{21}b_{11} + a_{22}b_{21})$$
$$= a_{11}b_{11}a_{21}b_{12} + a_{11}b_{11}a_{22}b_{22} + a_{12}b_{21}a_{21}b_{12} + a_{12}b_{21}a_{22}b_{22}$$
$$\quad - a_{11}b_{12}a_{21}b_{11} - a_{11}b_{12}a_{22}b_{21} - a_{12}b_{22}a_{21}b_{11} - a_{12}b_{22}a_{22}b_{21}$$
$$= a_{11}b_{11}a_{22}b_{22} + a_{12}b_{21}a_{21}b_{12} - a_{11}b_{12}a_{22}b_{21} - a_{12}b_{22}a_{21}b_{11}$$
$$= \det A \det B \quad \blacksquare$$

EXAMPLE 14 Verify Equation (8) for $A = \begin{pmatrix} 1 & -1 & 2 \\ 3 & 1 & 4 \\ 0 & -2 & 5 \end{pmatrix}$ and $B = \begin{pmatrix} 1 & -2 & 3 \\ 0 & -1 & 4 \\ 2 & 0 & -2 \end{pmatrix}$.

Solution $\det A = 16$ and $\det B = -8$. We calculate

$$AB = \begin{pmatrix} 1 & -1 & 2 \\ 3 & 1 & 4 \\ 0 & -2 & 5 \end{pmatrix} \begin{pmatrix} 1 & -2 & 3 \\ 0 & -1 & 4 \\ 2 & 0 & -2 \end{pmatrix} = \begin{pmatrix} 5 & -1 & -5 \\ 11 & -7 & 5 \\ 10 & 2 & -18 \end{pmatrix}$$

and $\det AB = -128 = (16)(-8) = \det A \det B$.

PROBLEMS 3.2 In Problems 1–20 evaluate the determinant by using the methods of this section.

1. $\begin{vmatrix} 3 & -5 \\ 2 & 6 \end{vmatrix}$

2. $\begin{vmatrix} 4 & 1 \\ 0 & -3 \end{vmatrix}$

3. $\begin{vmatrix} -1 & 0 & 2 \\ 3 & 1 & 4 \\ 2 & 0 & -6 \end{vmatrix}$

4. $\begin{vmatrix} 2 & 1 & -1 \\ 3 & -2 & 0 \\ 5 & 1 & 6 \end{vmatrix}$

5. $\begin{vmatrix} -3 & 2 & 4 \\ 1 & -1 & 2 \\ -1 & 4 & 0 \end{vmatrix}$

6. $\begin{vmatrix} 0 & -2 & 3 \\ 1 & 2 & -3 \\ 4 & 0 & 5 \end{vmatrix}$

7. $\begin{vmatrix} -2 & 3 & 6 \\ 4 & 1 & 8 \\ -2 & 0 & 0 \end{vmatrix}$

8. $\begin{vmatrix} 2 & -1 & 3 \\ 4 & 0 & 6 \\ 5 & -2 & 3 \end{vmatrix}$

9. $\begin{vmatrix} 1 & -1 & 2 & 4 \\ 0 & -3 & 5 & 6 \\ 1 & 4 & 0 & 3 \\ 0 & 5 & -6 & 7 \end{vmatrix}$

10. $\begin{vmatrix} 2 & -3 & 1 & 4 \\ 0 & -2 & 0 & 0 \\ 3 & 7 & -1 & 2 \\ 4 & 1 & -3 & 8 \end{vmatrix}$

11. $\begin{vmatrix} 1 & 1 & -1 & 0 \\ -3 & 4 & 6 & 0 \\ 2 & 5 & -1 & 3 \\ 4 & 0 & 3 & 0 \end{vmatrix}$

12. $\begin{vmatrix} 3 & -1 & 2 & 1 \\ 4 & 3 & 1 & -2 \\ -1 & 0 & 2 & 3 \\ 6 & 2 & 5 & 2 \end{vmatrix}$

13. $\begin{vmatrix} 2 & 0 & 0 & 0 \\ 0 & 0 & 3 & 0 \\ 0 & -1 & 0 & 0 \\ 0 & 0 & 0 & 4 \end{vmatrix}$

14. $\begin{vmatrix} 0 & a & 0 & 0 \\ b & 0 & 0 & 0 \\ 0 & 0 & 0 & c \\ 0 & 0 & d & 0 \end{vmatrix}$

15. $\begin{vmatrix} 1 & 2 & 0 & 0 \\ 3 & -2 & 0 & 0 \\ 0 & 0 & 1 & -5 \\ 0 & 0 & 7 & 2 \end{vmatrix}$

16. $\begin{vmatrix} a & b & 0 & 0 \\ c & d & 0 & 0 \\ 0 & 0 & a & -b \\ 0 & 0 & c & d \end{vmatrix}$

17. $\begin{vmatrix} 2 & -1 & 0 & 4 & 1 \\ 3 & 1 & -1 & 2 & 0 \\ 3 & 2 & -2 & 5 & 1 \\ 0 & 0 & 4 & -1 & 6 \\ 3 & 2 & 1 & -1 & 1 \end{vmatrix}$

18. $\begin{vmatrix} 1 & -1 & 2 & 0 & 0 \\ 3 & 1 & 4 & 0 & 0 \\ 2 & -1 & 5 & 0 & 0 \\ 0 & 0 & 0 & 2 & 3 \\ 0 & 0 & 0 & -1 & 4 \end{vmatrix}$

19. $\begin{vmatrix} a & 0 & 0 & 0 & 0 \\ 0 & 0 & b & 0 & 0 \\ 0 & 0 & 0 & 0 & c \\ 0 & 0 & 0 & d & 0 \\ 0 & e & 0 & 0 & 0 \end{vmatrix}$

20. $\begin{vmatrix} 2 & 5 & -6 & 8 & 0 \\ 0 & 1 & -7 & 6 & 0 \\ 0 & 0 & 0 & 4 & 0 \\ 0 & 2 & 1 & 5 & 1 \\ 4 & -1 & 5 & 3 & 0 \end{vmatrix}$

In Problems 21–27 compute the determinant assuming that

$$\begin{vmatrix} a_{11} & a_{12} & a_{13} \\ a_{21} & a_{22} & a_{23} \\ a_{31} & a_{32} & a_{33} \end{vmatrix} = 8$$

21. $\begin{vmatrix} a_{31} & a_{32} & a_{33} \\ a_{21} & a_{22} & a_{23} \\ a_{11} & a_{12} & a_{13} \end{vmatrix}$

22. $\begin{vmatrix} a_{31} & a_{32} & a_{33} \\ a_{11} & a_{12} & a_{13} \\ a_{21} & a_{22} & a_{23} \end{vmatrix}$

23. $\begin{vmatrix} a_{11} & a_{12} & a_{13} \\ 2a_{21} & 2a_{22} & 2a_{23} \\ a_{31} & a_{32} & a_{33} \end{vmatrix}$

24. $\begin{vmatrix} -3a_{11} & -3a_{12} & -3a_{13} \\ 2a_{21} & 2a_{22} & 2a_{23} \\ 5a_{31} & 5a_{32} & 5a_{33} \end{vmatrix}$

25. $\begin{vmatrix} a_{11} & 2a_{13} & a_{12} \\ a_{21} & 2a_{23} & a_{22} \\ a_{31} & 2a_{33} & a_{32} \end{vmatrix}$

26. $\begin{vmatrix} a_{11}-a_{12} & a_{12} & a_{13} \\ a_{21}-a_{22} & a_{22} & a_{23} \\ a_{31}-a_{32} & a_{32} & a_{33} \end{vmatrix}$

27. $\begin{vmatrix} 2a_{11}-3a_{21} & 2a_{12}-3a_{22} & 2a_{13}-3a_{23} \\ a_{31} & a_{32} & a_{33} \\ a_{21} & a_{22} & a_{23} \end{vmatrix}$

28. Using Property 2, show that if α is a number and A is an $n \times n$ matrix, then $\det \alpha A = \alpha^n \det A$.

*29. Show that

$$\begin{vmatrix} 1+x_1 & x_2 & x_3 & \cdots & x_n \\ x_1 & 1+x_2 & x_3 & \cdots & x_n \\ x_1 & x_2 & 1+x_3 & \cdots & x_n \\ \vdots & \vdots & \vdots & & \vdots \\ x_1 & x_2 & x_3 & \cdots & 1+x_n \end{vmatrix} = 1 + x_1 + x_2 + \cdots + x_n$$

*30. A matrix is **skew-symmetric** if $A^t = -A$. If A is an $n \times n$ skew-symmetric matrix, show that det $A = (-1)^n$ det A.

31. Using the result of Problem 30, show that if A is a skew-symmetric $n \times n$ matrix and n is odd, then det $A = 0$.

32. A matrix A is called **orthogonal** if A is invertible and $A^{-1} = A^t$. Show that if A is orthogonal, then det $A = \pm 1$.

**33. Let Δ denote the triangle in the plane with vertices at (x_1, y_1), (x_2, y_2), and (x_3, y_3). Show that the area of the triangle is given by

$$\text{Area of } \Delta = \pm \tfrac{1}{2} \begin{vmatrix} 1 & x_1 & y_1 \\ 1 & x_2 & y_2 \\ 1 & x_3 & y_3 \end{vmatrix}$$

Under what circumstances will this determinant equal zero?

**34. Three lines, no two of which are parallel, determine a triangle in the plane. Suppose that the lines are given by

$$a_{11}x + a_{12}y + a_{13} = 0$$
$$a_{21}x + a_{22}y + a_{23} = 0$$
$$a_{31}x + a_{32}y + a_{33} = 0$$

Show that the area determined by the lines is

$$\frac{\pm 1}{2A_{13}A_{23}A_{33}} \begin{vmatrix} A_{11} & A_{12} & A_{13} \\ A_{21} & A_{22} & A_{23} \\ A_{31} & A_{32} & A_{33} \end{vmatrix}$$

35. The 3×3 Vandermonde† determinant is given by

$$D_3 = \begin{vmatrix} 1 & 1 & 1 \\ a_1 & a_2 & a_3 \\ a_1^2 & a_2^2 & a_3^2 \end{vmatrix}$$

Show that $D_3 = (a_2 - a_1)(a_3 - a_1)(a_3 - a_2)$.

36. $D_4 = \begin{vmatrix} 1 & 1 & 1 & 1 \\ a_1 & a_2 & a_3 & a_4 \\ a_1^2 & a_2^2 & a_3^2 & a_4^2 \\ a_1^3 & a_2^3 & a_3^3 & a_4^3 \end{vmatrix}$ is the 4×4 Vandermonde determinant.†

Show that $D_4 = (a_2 - a_1)(a_3 - a_1)(a_4 - a_1)(a_3 - a_2)(a_4 - a_2)(a_4 - a_3)$.

**37. a. Define the $n \times n$ Vandermonde determinant D_n.

---

† A. T. Vandermonde (1735–1796) was a French mathematician.

b. Show that $D_n = \prod_{\substack{i=1 \\ j>i}}^{n} (a_j - a_i)$, where Π stands for the word "product." Note that the product in Problem 36 can be written $\prod_{\substack{i=1 \\ j>i}}^{4} (a_j - a_i)$.

3.3 If Time Permits: Proof of the Basic Theorem

THEOREM 1

BASIC THEOREM Let $A = (a_{ij})$ be an $n \times n$ matrix. Then

$$\det A = a_{11}A_{11} + a_{12}A_{12} + \cdots + a_{1n}A_{1n}$$
$$= a_{i1}A_{i1} + a_{i2}A_{i2} + \cdots + a_{in}A_{in} \tag{1}$$
$$= a_{1j}A_{1j} + a_{2j}A_{2j} + \cdots + a_{nj}A_{nj} \tag{2}$$

for $i = 1, 2, \ldots, n$ and $j = 1, 2, \ldots, n$.

Note. The first equality is Definition 3.1.3 of the determinant by cofactor expansion in the first row; the second equality says that the expansion by cofactors in any other row yields the determinant; the third equality says that expansion by cofactors in any column gives the determinant.

Proof We prove equality (1) by mathematical induction. For the 2×2 matrix $A = \begin{pmatrix} a_{11} & a_{12} \\ a_{21} & a_{22} \end{pmatrix}$, we first expand the first row by cofactors: $\det A = a_{11}A_{11} + a_{12}A_{12} = a_{11}(a_{22}) + a_{12}(-a_{21}) = a_{11}a_{22} - a_{12}a_{21}$. Similarly, expanding in the second row, we obtain $a_{21}A_{21} + a_{22}A_{22} = a_{21}(-a_{12}) + a_{22}(a_{11}) = a_{11}a_{22} - a_{12}a_{21}$. Thus we get the same result by expanding in any row of a 2×2 matrix and this proves equality (1) in the 2×2 case.

We now assume that equality (1) holds for all $(n-1) \times (n-1)$ matrices. We must show that it holds for $n \times n$ matrices. Our procedure will be to expand by cofactors in the first and ith rows and show that the expansions are identical. If we expand in the first row, then a typical term in the cofactor expansion is

$$a_{1k}A_{1k} = (-1)^{1+k}a_{1k}|M_{1k}| \tag{3}$$

Note that this is the only place in the expansion of $|A|$ that the term a_{1k} occurs since another typical term is $a_{1m}A_{1m} = (-1)^{1+m}|M_{1m}|$, $k \neq m$, and M_{1m} is obtained by deleting the first row and mth column of A (and a_{1k} is in the first row of A). Since M_{1k} is an $(n-1) \times (n-1)$ matrix, we can, by the induction hypothesis, calculate $|M_{1k}|$ by expanding in the ith row of A (which is the $(i-1)$st row of M_{1k}). A typical term in this expansion is

$$a_{il} \text{ (cofactor of } a_{il} \text{ in } M_{1k}) \quad (k \neq l) \tag{4}$$

For the reasons outlined above, this is the only term in the expansion of $|M_{1k}|$ in the ith row of A that contains the term a_{il}. Substituting (4) into (3), we find that

$$(-1)^{1+k}a_{1k}a_{il} \text{ (cofactor of } a_{il} \text{ in } M_{1k}) \quad (k \neq l) \tag{5}$$

is the only occurrence of the term $a_{1k}a_{il}$ in the cofactor expansion of det A in the first row.

Now if we expand by cofactors in the ith row of A (where $i \neq 1$), a typical term is

$$(-1)^{i+1} a_{il} |M_{il}| \tag{6}$$

and a typical term in the expansion of $|M_{il}|$ in the first row of M_{il} is

$$a_{1k} \text{ (cofactor of } a_{1k} \text{ in } M_{il}) \quad (k \neq l) \tag{7}$$

and, inserting (7) in (6), we find that the only occurrence of the term $a_{il}a_{1k}$ in the expansion of det A along its ith row is

$$(-1)^{i+1} a_{1k} a_{il} \text{ (cofactor of } a_{1k} \text{ in } M_{il}) \quad (k \neq l) \tag{8}$$

If we can show that the expressions in (5) and (8) are the same, then (1) will be proved, for the term in (5) is the only occurrence of $a_{1k}a_{il}$ in the first row expansion, the term in (8) is the only occurrence of $a_{1k}a_{il}$ in the ith row expansion, and k, i, and l are arbitrary. This will show that the sums of the terms in the first and ith row expansions are the same.

Now let $M_{1i,kl}$ denote the $(n-2) \times (n-2)$ matrix obtained by deleting the first and ith rows and kth and lth columns of A. (This is called a **second-order minor** of A.) We first suppose that $k < l$. Then

$$M_{1k} = \begin{pmatrix} a_{21} & \cdots & a_{2,k-1} & a_{2,k+1} & \cdots & a_{2l} & \cdots & a_{2n} \\ \vdots & & \vdots & \vdots & & \vdots & & \vdots \\ a_{i1} & \cdots & a_{i,k-1} & a_{i,k+1} & \cdots & a_{il} & \cdots & a_{in} \\ \vdots & & \vdots & \vdots & & \vdots & & \vdots \\ a_{n1} & \cdots & a_{n,k-1} & a_{n,k+1} & \cdots & a_{nl} & \cdots & a_{nn} \end{pmatrix} \tag{9}$$

$$M_{il} = \begin{pmatrix} a_{11} & \cdots & a_{1k} & \cdots & a_{1,l-1} & a_{1,l+1} & \cdots & a_{1n} \\ \vdots & \ddots & \vdots & & \vdots & \vdots & & \vdots \\ a_{i-1,1} & \cdots & a_{i-1,k} & \cdots & a_{i-1,l-1} & a_{i-1,l+1} & \cdots & a_{i-1,n} \\ a_{i+1,1} & \cdots & a_{i+1,k} & \cdots & a_{i+1,l-1} & a_{i+1,l+1} & \cdots & a_{i+1,n} \\ \vdots & & \vdots & & \vdots & \vdots & & \vdots \\ a_{n1} & \cdots & a_{nk} & \cdots & a_{n,l-1} & a_{n,l+1} & \cdots & a_{nn} \end{pmatrix} \tag{10}$$

From (9) and (10), we see that

$$\text{Cofactor of } a_{il} \text{ in } M_{1k} = (-1)^{(i-1)+(l-1)} |M_{1i,kl}| \tag{11}$$

$$\text{Cofactor of } a_{1k} \text{ in } M_{il} = (-1)^{1+k} |M_{1i,kl}| \tag{12}$$

Thus (5) becomes

$$(-1)^{1+k} a_{1k} a_{il} (-1)^{(i-1)+(l-1)} |M_{1i,kl}| = (-1)^{i+k+l-1} a_{1k} a_{il} |M_{1i,kl}| \tag{13}$$

and (8) becomes

$$(-1)^{i+l} a_{1k} a_{il} (-1)^{1+k} |M_{1i,kl}| = (-1)^{i+k+l+1} a_{1k} a_{il} |M_{1i,kl}| \tag{14}$$

But $(-1)^{i+k+l-1} = (-1)^{i+k+l+1}$, so the right sides of Equations (13) and (14) are equal. Hence expressions (5) and (8) are equal and (1) is proved in the case $k<l$. If $k>l$, then, by similar reasoning, we find that

$$\text{Cofactor of } a_{il} \text{ in } M_{1k} = (-1)^{(i-1)+l}|M_{1i,kl}|$$

$$\text{Cofactor of } a_{1k} \text{ in } M_{il} = (-1)^{1+(k-1)}|M_{1i,kl}|$$

so that (5) becomes

$$(-1)^{1+k}a_{1k}a_{il}(-1)^{(i-1)+l}|M_{1i,kl}| = (-1)^{i+k+l}a_{1k}a_{il}|M_{1i,kl}|$$

and (8) becomes

$$(-1)^{i+l}a_{1k}a_{il}(-1)^{1+k}|M_{1i,kl}| = (-1)^{i+k+l+1}a_{1k}a_{il}|M_{1i,kl}|$$

This completes the proof of Equation (1).

To prove Equation (2) we go through a similar process. If we expand in the kth and lth columns, we find that the only occurrences of the term $a_{1k}a_{il}$ will be given by (5) and (8). (See Problems 1 and 2.) This shows that the expansion by cofactors in any two columns is the same and that each is equal to the expansion along any row. This completes the proof. ■

PROBLEMS 3.3

1. Show that if A is expanded along its kth column, then the only occurrence of the term $a_{1k}a_{il}$ is given by Equation (5).
2. Show that if A is expanded along its lth column, then the only occurrence of the term $a_{1k}a_{il}$ is given by Equation (8).
3. Show that if A is expanded along its kth column, then the only occurrence of the term $a_{ik}a_{jl}$ is $(-1)^{i+k}a_{ik}a_{jl}$ (cofactor of a_{jl} in M_{ik}) for $l \neq k$.
4. Let $A = \begin{pmatrix} 1 & 5 & 7 \\ 2 & -1 & 3 \\ 4 & 5 & -2 \end{pmatrix}$. Compute det A by expanding in each of the rows and columns.
5. Do the same for the matrix $A = \begin{pmatrix} 1 & -1 & 4 \\ 0 & 1 & 5 \\ -3 & 7 & 2 \end{pmatrix}$.

3.4 Determinants and Inverses

In this section we shall see how matrix inverses can be calculated by using determinants. Moreover, we shall complete the task, begun in Chapter 1, of proving the important Summing Up Theorem 2.7.6 (see page 72) showing the equivalence of various properties of matrices. We begin with a simple result.

THEOREM 1 If A is invertible, then det $A \neq 0$ and

$$\det A^{-1} = \frac{1}{\det A} \tag{1}$$

Proof From Theorems 3.2.4, page 99, and 3.1.1, page 87, we have

$$1 = \det I = \det AA^{-1} = \det A \det A^{-1} \blacksquare \tag{2}$$

If det A were equal to zero, then Equation (2) would read $1 = 0$. Thus det $A \neq 0$ and det $A^{-1} = 1/\det A$.

Before using determinants to calculate inverses, we need to define the *adjoint* of a matrix $A = (a_{ij})$. Let $B = (A_{ij})$ be the matrix of cofactors of A. (Remember that a cofactor, defined on page 85, is a number.) Then

$$B = \begin{pmatrix} A_{11} & A_{12} & \cdots & A_{1n} \\ A_{21} & A_{22} & \cdots & A_{2n} \\ \vdots & \vdots & & \vdots \\ A_{n1} & A_{n2} & \cdots & A_{nn} \end{pmatrix} \tag{3}$$

DEFINITION 1 **THE ADJOINT** Let A be an $n \times n$ matrix and let B, given by (3), denote the matrix of its cofactors. Then the **adjoint** *of* A, written adj A, is the transpose of the $n \times n$ matrix B; that is,

$$\operatorname{adj} A = B^t = \begin{pmatrix} A_{11} & A_{21} & \cdots & A_{n1} \\ A_{12} & A_{22} & \cdots & A_{n2} \\ \vdots & \vdots & & \vdots \\ A_{1n} & A_{2n} & \cdots & A_{nn} \end{pmatrix} \tag{4}$$

EXAMPLE 1 Let $A = \begin{pmatrix} 2 & 4 & 3 \\ 0 & 1 & -1 \\ 3 & 5 & 7 \end{pmatrix}$. Compute adj A.

Solution We have $A_{11} = \begin{vmatrix} 1 & -1 \\ 5 & 7 \end{vmatrix} = 12$, $A_{12} = -\begin{vmatrix} 0 & -1 \\ 3 & 7 \end{vmatrix} = -3$, $A_{13} = -3$, $A_{21} = -13$, $A_{22} = 5$, $A_{23} = 2$, $A_{31} = -7$, $A_{32} = 2$, and $A_{33} = 2$. Thus $B = \begin{pmatrix} 12 & -3 & -3 \\ -13 & 5 & 2 \\ -7 & 2 & 2 \end{pmatrix}$ and adj $A = B^t = \begin{pmatrix} 12 & -13 & -7 \\ -3 & 5 & 2 \\ -3 & 2 & 2 \end{pmatrix}$.

EXAMPLE 2 Let

$$A = \begin{pmatrix} 1 & -3 & 0 & -2 \\ 3 & -12 & -2 & -6 \\ -2 & 10 & 2 & 5 \\ -1 & 6 & 1 & 3 \end{pmatrix}$$

Calculate adj A.

Solution This is more tedious since we have to compute sixteen 3×3 determinants. For example, we have $A_{12} = -\begin{vmatrix} 3 & -2 & -6 \\ -2 & 2 & 5 \\ -1 & 1 & 3 \end{vmatrix} = -1$, $A_{24} = \begin{vmatrix} 1 & -3 & 0 \\ -2 & 10 & 2 \\ -1 & 6 & 1 \end{vmatrix} = -2$, and $A_{43} = -\begin{vmatrix} 1 & -3 & -2 \\ 3 & -12 & -6 \\ -2 & 10 & 5 \end{vmatrix} = 3$. Completing these calculations, we find that

$$B = \begin{pmatrix} 0 & -1 & 0 & 2 \\ -1 & 1 & -1 & -2 \\ 0 & 2 & -3 & -3 \\ -2 & -2 & 3 & 2 \end{pmatrix}$$

$$\text{adj } A = B^t = \begin{pmatrix} 0 & -1 & 0 & -2 \\ -1 & 1 & 2 & -2 \\ 0 & -1 & -3 & 3 \\ 2 & -2 & -3 & 2 \end{pmatrix}$$

EXAMPLE 3 Let $A = \begin{pmatrix} a_{11} & a_{12} \\ a_{21} & a_{22} \end{pmatrix}$. Then adj $A = \begin{pmatrix} A_{11} & A_{21} \\ A_{12} & A_{22} \end{pmatrix} = \begin{pmatrix} a_{22} & -a_{12} \\ -a_{21} & a_{11} \end{pmatrix}$.

Warning. In taking the adjoint of a matrix, do not forget to transpose the matrix of cofactors.

THEOREM 2 Let A be an $n \times n$ matrix. Then

$$(A)(\text{adj } A) = \begin{pmatrix} \det A & 0 & 0 & \cdots & 0 \\ 0 & \det A & 0 & \cdots & 0 \\ 0 & 0 & \det A & \cdots & 0 \\ \vdots & \vdots & \vdots & & \vdots \\ 0 & 0 & 0 & \cdots & \det A \end{pmatrix} = (\det A)I \quad (5)$$

Proof Let $C = (c_{ij}) = (A)(\text{adj } A)$. Then

$$C = \begin{pmatrix} a_{11} & a_{12} & \cdots & a_{1n} \\ a_{21} & a_{22} & \cdots & a_{2n} \\ \vdots & \vdots & & \vdots \\ a_{n1} & a_{n2} & \cdots & a_{nn} \end{pmatrix} \begin{pmatrix} A_{11} & A_{21} & \cdots & A_{n1} \\ A_{12} & A_{22} & \cdots & A_{n2} \\ \vdots & \vdots & & \vdots \\ A_{1n} & A_{2n} & \cdots & A_{nn} \end{pmatrix} \quad (6)$$

We have

$$c_{ij} = (i\text{th row of } A) \cdot (j\text{th column of adj } A)$$

$$= (a_{i1} \quad a_{i2} \cdots a_{in}) \cdot \begin{pmatrix} A_{j1} \\ A_{j2} \\ \vdots \\ A_{jn} \end{pmatrix}$$

Thus
$$c_{ij} = a_{i1}A_{j1} + a_{i2}A_{j2} + \cdots + a_{in}A_{jn} \quad (7)$$

Now if $i = j$, the sum in (7) equals $a_{i1}A_{i1} + a_{i2}A_{i2} + \cdots + a_{in}A_{in}$, which is the expansion of det A in the ith row of A. On the other hand, if $i \neq j$, then from Theorem 3.2.2 on page 97, the sum in (7) equals zero. Thus

$$c_{ij} = \begin{cases} \det A & \text{if } i = j \\ 0 & \text{if } i \neq j \end{cases}$$

This proves the theorem. ∎

We can now state the main result.

THEOREM 3 Let A be an $n \times n$ matrix. Then A is invertible if and only if det $A \neq 0$. If det $A \neq 0$, then

$$A^{-1} = \frac{1}{\det A} \text{adj } A \quad (8)$$

Note that Theorem 2.7.4 on page 65 for 2×2 matrices is a special case of this theorem.

Proof If A is invertible, then det $A \neq 0$ by Theorem 1. If det $A \neq 0$, then,

$$(A)\left(\frac{1}{\det A} \text{adj } A\right) = \frac{1}{\det A}[A(\text{adj } A)] \overset{\text{Theorem 2}}{=} \frac{1}{\det A}(\det A)I = I$$

But, by Theorem 2.7.8 on page 74, if $AB = I$, then $B = A^{-1}$. Thus

$$(1/\det A) \text{adj } A = A^{-1}. \blacksquare$$

EXAMPLE 4 Let $A = \begin{pmatrix} 2 & 4 & 3 \\ 0 & 1 & -1 \\ 3 & 5 & 7 \end{pmatrix}$. Determine whether A is invertible and calculate A^{-1} if it is.

Solution Since $\det A = 3 \neq 0$, we see that A is invertible. From Example 1,
$$\operatorname{adj} A = \begin{pmatrix} 12 & -13 & -7 \\ -3 & 5 & 2 \\ -3 & 2 & 2 \end{pmatrix}.$$

Thus
$$A^{-1} = \tfrac{1}{3} \begin{pmatrix} 12 & -13 & -7 \\ -3 & 5 & 2 \\ -3 & 2 & 2 \end{pmatrix} = \begin{pmatrix} 4 & -\tfrac{13}{3} & -\tfrac{7}{3} \\ -1 & \tfrac{5}{3} & \tfrac{2}{3} \\ -1 & \tfrac{2}{3} & \tfrac{2}{3} \end{pmatrix}$$

Check. $A^{-1} A = \tfrac{1}{3} \begin{pmatrix} 12 & -13 & -7 \\ -3 & 5 & 2 \\ -3 & 2 & 2 \end{pmatrix} \begin{pmatrix} 2 & 4 & 3 \\ 0 & 1 & -1 \\ 3 & 5 & 7 \end{pmatrix} = \tfrac{1}{3} \begin{pmatrix} 3 & 0 & 0 \\ 0 & 3 & 0 \\ 0 & 0 & 3 \end{pmatrix} = I$

EXAMPLE 5 Let
$$A = \begin{pmatrix} 1 & -3 & 0 & -2 \\ 3 & -12 & -2 & -6 \\ -2 & 10 & 2 & 5 \\ -1 & 6 & 1 & 3 \end{pmatrix}$$

Determine whether A is invertible and, if so, calculate A^{-1}.

Solution Using properties of determinants, we compute
$$\begin{vmatrix} 1 & -3 & 0 & -2 \\ 3 & -12 & -2 & -6 \\ -2 & 10 & 2 & 5 \\ -1 & 6 & 1 & 3 \end{vmatrix}$$

Multiply the first column by 3 and 2 and add it to the second and fourth columns, respectively.
$$= \begin{vmatrix} 1 & 0 & 0 & 0 \\ 3 & -3 & -2 & 0 \\ -2 & 4 & 2 & 1 \\ -1 & 3 & 1 & 1 \end{vmatrix}$$

Expand in the first row.
$$= \begin{vmatrix} -3 & -2 & 0 \\ 4 & 2 & 1 \\ 3 & 1 & 1 \end{vmatrix} = -1$$

Thus $\det A = -1 \neq 0$ and A^{-1} exists. By Example 2, we have
$$\operatorname{adj} A = \begin{pmatrix} 0 & -1 & 0 & -2 \\ -1 & 1 & 2 & -2 \\ 0 & -1 & -3 & 3 \\ 2 & -2 & -3 & 3 \end{pmatrix}$$

Thus
$$A^{-1} = \frac{1}{-1} \begin{pmatrix} 0 & -1 & 0 & -2 \\ -1 & 1 & 2 & -2 \\ 0 & -1 & -3 & 3 \\ 2 & -2 & -3 & 2 \end{pmatrix} = \begin{pmatrix} 0 & 1 & 0 & 2 \\ 1 & -1 & -2 & 2 \\ 0 & 1 & 3 & -3 \\ -2 & 2 & 3 & -2 \end{pmatrix}$$

Check. $AA^{-1} = \begin{pmatrix} 1 & -3 & 0 & -2 \\ 3 & -12 & -2 & -6 \\ -2 & 10 & 2 & 5 \\ -1 & 6 & 1 & 3 \end{pmatrix} \begin{pmatrix} 0 & 1 & 0 & 2 \\ 1 & -1 & -2 & 2 \\ 0 & 1 & 3 & -3 \\ -2 & 2 & 3 & -2 \end{pmatrix}$

$= \begin{pmatrix} 1 & 0 & 0 & 0 \\ 0 & 1 & 0 & 0 \\ 0 & 0 & 1 & 0 \\ 0 & 0 & 0 & 1 \end{pmatrix}$

Note. As you may have noticed, if $n > 3$ it is generally easier to compute A^{-1} by row reduction then by using adj A since, even for the 4×4 case, it is necessary to calculate 17 determinants (16 for the adjoint plus det A). Nevertheless, Theorem 3 is very important since, before you do any row reduction, the calculation of det A (if it can be done easily) will tell you whether or not A^{-1} exists.

We last saw out Summing Up Theorem (Theorems 1.2.1, page 4, and 2.7.6, page 72) in Section 2.7. This is the theorem that ties together many of the concepts developed in the first three chapters of this book. We are now able to prove the last two parts of that theorem.

THEOREM 4 **SUMMING UP THEOREM—VIEW 3** Let A be an $n \times n$ matrix. Then each of the following six statements implies the other five. (That is, if one is true, all are true.)

 i. A is invertible.
 ii. The only solution to the homogeneous system $A\mathbf{x} = \mathbf{0}$ is the trivial solution ($\mathbf{x} = \mathbf{0}$).
 iii. The system $A\mathbf{x} = \mathbf{b}$ has a unique solution for every n-vector $\mathbf{b}$.
 iv. A is row equivalent to the $n \times n$ identity matrix I_n.
 v. The rows (and columns) of A are linearly independent.
 vi. det $A \neq 0$.

Proof In Theorem 2.7.6 we proved the equivalence of parts (i), (ii), (iii), and (iv). In Theorem 3 of this section we saw the equivalence of (i) and (vi). To complete the proof we shall show that (v) is equivalent to (vi). Theorem 2.6.3 on page 57 states that the columns of A are linearly dependent if and only if the system $A\mathbf{x} = \mathbf{0}$ has an infinite number of solutions. This implies that the columns of A are linearly independent if and only if the system $A\mathbf{x} = \mathbf{0}$ has the unique solution $\mathbf{x} = \mathbf{0}$. Thus part (ii) is equivalent to part (v) for columns. Since parts (ii) and (vi) are equivalent, this means that the columns of A are linearly independent if and only if det $A \neq 0$. But the rows of A are the columns of A^t and, by Theorem 3.2.3 on page 98, det $A = $ det A^t. Thus the rows of A are linearly independent if and only if det $A^t = $ det $A \neq 0$. This means that parts (v) and (vi) are equivalent and the proof is complete. ∎

EXAMPLE 6

Determine whether the vectors $\begin{pmatrix} 1 \\ -2 \\ 3 \end{pmatrix}$, $\begin{pmatrix} 4 \\ 1 \\ 5 \end{pmatrix}$, and $\begin{pmatrix} 1 \\ 0 \\ 2 \end{pmatrix}$ are linearly independent or dependent.

Solution Let $A = \begin{pmatrix} 1 & 4 & 1 \\ -2 & 1 & 0 \\ 3 & 5 & 2 \end{pmatrix}$. Then $|A| = \begin{vmatrix} 1 & 4 & 1 \\ -2 & 1 & 0 \\ 3 & 5 & 2 \end{vmatrix} = 5 \neq 0$, and hence the vectors (which are the columns of A) are linearly independent.

EXAMPLE 7

Determine whether the vectors $\begin{pmatrix} -2 \\ 4 \\ 5 \end{pmatrix}$, $\begin{pmatrix} 3 \\ 1 \\ 0 \end{pmatrix}$, and $\begin{pmatrix} 4 \\ 6 \\ 5 \end{pmatrix}$ are linearly independent or dependent.

Solution Proceeding as in Example 6, we have $\begin{vmatrix} -2 & 3 & 4 \\ 4 & 1 & 6 \\ 5 & 0 & 5 \end{vmatrix} = 0$; thus the vectors are linearly dependent.

EXAMPLE 8

Determine whether the vectors $(-1, 0, 4)$, $(2, 3, 1)$, and $(5, 2, 0)$ are linearly independent or dependent.

Solution Let $A = \begin{pmatrix} -1 & 0 & 4 \\ 2 & 3 & 1 \\ 5 & 2 & 0 \end{pmatrix}$. Since $\det A = -42 \neq 0$, we conclude that the rows of A are linearly independent.

PROBLEMS 3.4

In Problems 1–12 use the methods of this section to determine whether the given matrix is invertible. If so, compute the inverse.

1. $\begin{pmatrix} 3 & 2 \\ 1 & 2 \end{pmatrix}$

2. $\begin{pmatrix} 3 & 6 \\ -4 & -8 \end{pmatrix}$

3. $\begin{pmatrix} 0 & 1 \\ 1 & 0 \end{pmatrix}$

4. $\begin{pmatrix} 1 & 1 & 1 \\ 0 & 2 & 3 \\ 5 & 5 & 1 \end{pmatrix}$

5. $\begin{pmatrix} 3 & 2 & 1 \\ 0 & 2 & 2 \\ 0 & 1 & -1 \end{pmatrix}$

6. $\begin{pmatrix} 1 & 1 & 1 \\ 0 & 1 & 1 \\ 0 & 0 & 1 \end{pmatrix}$

7. $\begin{pmatrix} 1 & 2 & 3 \\ 1 & 1 & 2 \\ 0 & 1 & 2 \end{pmatrix}$

8. $\begin{pmatrix} 3 & 1 & 0 \\ 1 & -1 & 2 \\ 1 & 1 & 1 \end{pmatrix}$

9. $\begin{pmatrix} 2 & -1 & 4 \\ -1 & 0 & 5 \\ 19 & -7 & 3 \end{pmatrix}$

10. $\begin{pmatrix} 1 & 6 & 2 \\ -2 & 3 & 5 \\ 7 & 12 & -4 \end{pmatrix}$

11. $\begin{pmatrix} 1 & 1 & 1 & 1 \\ 1 & 2 & -1 & 2 \\ 1 & -1 & 2 & 1 \\ 1 & 3 & 3 & 2 \end{pmatrix}$

12. $\begin{pmatrix} 1 & -3 & 0 & -2 \\ 3 & -12 & -2 & -6 \\ -2 & 10 & 2 & 5 \\ -1 & 6 & 1 & 3 \end{pmatrix}$

13. Using determinants, determine whether the vectors $\begin{pmatrix} 2 \\ -1 \\ 4 \end{pmatrix}, \begin{pmatrix} 1 \\ 6 \\ 2 \end{pmatrix}, \begin{pmatrix} 0 \\ 0 \\ 1 \end{pmatrix}$ are linearly dependent or independent.

14. Do the same for the vectors $\begin{pmatrix} 1 \\ -3 \\ 2 \\ 1 \end{pmatrix}, \begin{pmatrix} 4 \\ 0 \\ 1 \\ 2 \end{pmatrix}, \begin{pmatrix} 0 \\ 3 \\ -1 \\ 4 \end{pmatrix}, \begin{pmatrix} 5 \\ -6 \\ 4 \\ -1 \end{pmatrix}$.

15. Do the same for the vectors $\begin{pmatrix} 3 \\ 0 \\ 1 \\ 0 \end{pmatrix}, \begin{pmatrix} 5 \\ 1 \\ -1 \\ 2 \end{pmatrix}, \begin{pmatrix} 6 \\ 0 \\ 1 \\ 0 \end{pmatrix}, \begin{pmatrix} 0 \\ 0 \\ 1 \\ 1 \end{pmatrix}$.

16. Show that the n n-vectors $\begin{pmatrix} 1 \\ 0 \\ 0 \\ \vdots \\ 0 \end{pmatrix}, \begin{pmatrix} 0 \\ 1 \\ 0 \\ \vdots \\ 0 \end{pmatrix}, \begin{pmatrix} 0 \\ 0 \\ 1 \\ 0 \\ \vdots \\ 0 \end{pmatrix}, \ldots, \begin{pmatrix} 0 \\ 0 \\ 0 \\ \vdots \\ 1 \\ 0 \end{pmatrix}, \begin{pmatrix} 0 \\ 0 \\ 0 \\ \vdots \\ 1 \end{pmatrix}$ are linearly independent.

17. Show that an $n \times n$ matrix A is invertible if and only if A^t is invertible.

18. For $A = \begin{pmatrix} 1 & 1 \\ 2 & 5 \end{pmatrix}$, verify that $\det A^{-1} = 1/\det A$.

19. For $A = \begin{pmatrix} 1 & -1 & 3 \\ 4 & 1 & 6 \\ 2 & 0 & -2 \end{pmatrix}$, verify that $\det A^{-1} = 1/\det A$.

20. For what values of α is the matrix $\begin{pmatrix} \alpha & -3 \\ 4 & 1-\alpha \end{pmatrix}$ not invertible?

21. For what values of α does the matrix $\begin{pmatrix} -\alpha & \alpha-1 & \alpha+1 \\ 1 & 2 & 3 \\ 2-\alpha & \alpha+3 & \alpha+7 \end{pmatrix}$ not have an inverse?

22. Suppose that the $n \times n$ matrix A is not invertible. Show that $(A)(\operatorname{adj} A)$ is the zero matrix.

3.5 Cramer's Rule

In this section we examine an old method for solving systems with the same number of unknowns as equations. Consider the system of n equations in n unknowns

$$\begin{aligned} a_{11}x_1 + a_{12}x_2 + \cdots + a_{1n}x_n &= b_1 \\ a_{21}x_1 + a_{22}x_2 + \cdots + a_{2n}x_n &= b_2 \\ &\vdots \\ a_{n1}x_1 + a_{n2}x_2 + \cdots + a_{nn}x_n &= b_n \end{aligned} \qquad (1)$$

which can be written in the form

$$A\mathbf{x} = \mathbf{b} \qquad (2)$$

We suppose that $\det A \neq 0$. Then system (2) has a unique solution given by $\mathbf{x} = A^{-1}\mathbf{b}$. We can develop a method for finding that solution without row reduction and without computing A^{-1}.

Let $D = \det A$. We define n new matrices:

$$A_1 = \begin{pmatrix} b_1 & a_{12} & \cdots & a_{1n} \\ b_2 & a_{22} & \cdots & a_{2n} \\ \vdots & \vdots & & \vdots \\ b_n & a_{n2} & \cdots & a_{nn} \end{pmatrix}, A_2 = \begin{pmatrix} a_{11} & b_1 & \cdots & a_{1n} \\ a_{21} & b_2 & \cdots & a_{2n} \\ \vdots & \vdots & & \vdots \\ a_{n1} & b_n & \cdots & a_{nn} \end{pmatrix}, \ldots,$$

$$A_n = \begin{pmatrix} a_{11} & a_{12} & \cdots & b_1 \\ a_{21} & a_{22} & \cdots & b_2 \\ \vdots & \vdots & & \vdots \\ a_{n1} & a_{n2} & \cdots & b_n \end{pmatrix}$$

That is, A_i is the matrix obtained by replacing the ith column of A with $\mathbf{b}$. Finally, let $D_1 = \det A_1, D_2 = \det A_2, \ldots, D_n = \det A_n$.

THEOREM 1   **CRAMER'S RULE**[†] Let A be an $n \times n$ matrix and suppose that $\det A \neq 0$. Then the unique solution to the system $A\mathbf{x} = \mathbf{b}$ is given by

$$x_1 = \frac{D_1}{D}, \ x_2 = \frac{D_2}{D}, \ \ldots, \ x_i = \frac{D_i}{D}, \ \ldots, \ x_n = \frac{D_n}{D} \tag{3}$$

Proof   The solution to $A\mathbf{x} = \mathbf{b}$ is $\mathbf{x} = A^{-1}\mathbf{b}$. But

$$A^{-1}\mathbf{b} = \frac{1}{D}(\text{adj } A)\mathbf{b} = \frac{1}{D} \begin{pmatrix} A_{11} & A_{21} & \cdots & A_{n1} \\ A_{12} & A_{22} & \cdots & A_{n2} \\ \vdots & \vdots & & \vdots \\ A_{1n} & A_{2n} & \cdots & A_{nn} \end{pmatrix} \begin{pmatrix} b_1 \\ b_2 \\ \vdots \\ b_n \end{pmatrix} \tag{4}$$

Now $(\text{adj } A)\mathbf{b}$ is an n-vector, the jth component of which is

$$(A_{1j} \ A_{2j} \ \cdots \ A_{nj}) \cdot \begin{pmatrix} b_1 \\ b_2 \\ \vdots \\ b_n \end{pmatrix} = b_1 A_{1j} + b_2 A_{2j} + \cdots + b_n A_{nj} \tag{5}$$

[†] Named for the Swiss mathematician Gabriel Cramer (1704–1752). Cramer published the rule in 1750 in his *Introduction to the Analysis of Lines of Algebraic Curves*. Actually, there is much evidence to suggest that the rule was known as early as 1729 to Colin Maclaurin (1698–1746), who was probably the most outstanding British mathematician in the years following the death of Newton.

Consider the matrix A_j:

$$A_j = \begin{pmatrix} a_{11} & a_{12} & \cdots & b_1 & \cdots & a_{1n} \\ a_{21} & a_{22} & \cdots & b_2 & \cdots & a_{2n} \\ \vdots & \vdots & & \vdots & & \vdots \\ a_{n1} & a_{n2} & \cdots & b_n & \cdots & a_{nn} \end{pmatrix} \qquad (6)$$

$$\uparrow$$
$$j\text{th column}$$

If we expand the determinant of A_j in its jth column, we obtain

$$D_j = b_1 \text{ (cofactor of } b_1) + b_2 \text{ (cofactor of } b_2) + \cdots \\ + b_n \text{ (cofactor of } b_n) \qquad (7)$$

But to find the cofactor of b_i, say, we delete the ith row and jth column of A_j (since b_i is in the jth column of A_j). But the jth column of A_j is $\mathbf{b}$ and, with this deleted, we simply have the ij minor, M_{ij}, of A. Thus

$$\text{Cofactor of } b_i \text{ in } A_j = A_{ij}$$

so that (7) becomes

$$D_j = b_1 A_{1j} + b_2 A_{2j} + \cdots + b_n A_{nj} \qquad (8)$$

But this is the same as the right side of (5). Thus the ith component of $(\operatorname{adj} A)\mathbf{b}$ is D_i and we have

$$\mathbf{x} = \begin{pmatrix} x_1 \\ x_2 \\ \vdots \\ x_n \end{pmatrix} = A^{-1}\mathbf{b} = \frac{1}{D}(\operatorname{adj} A)\mathbf{b} = \frac{1}{D}\begin{pmatrix} D_1 \\ D_2 \\ \vdots \\ D_n \end{pmatrix} = \begin{pmatrix} D_1/D \\ D_2/D \\ \vdots \\ D_n/D \end{pmatrix}$$

and the proof is complete. ∎

EXAMPLE 1 Solve, using Cramer's rule, the system

$$\begin{aligned} 2x_1 + 4x_2 + 6x_3 &= 18 \\ 4x_1 + 5x_2 + 6x_3 &= 24 \\ 3x_1 + x_2 - 2x_3 &= 4 \end{aligned} \qquad (9)$$

Solution We have solved this before—using row reduction in Example 1.3.1 on page 6. We could also solve it by calculating A^{-1} (Example 2.7.6, page 66) and then finding $A^{-1}\mathbf{b}$. We now solve it by using Cramer's rule. First we have

$$D = \begin{vmatrix} 2 & 4 & 6 \\ 4 & 5 & 6 \\ 3 & 1 & -2 \end{vmatrix} = 6 \neq 0.$$

so that system (9) has a unique solution. Then $D_1 = \begin{vmatrix} 18 & 4 & 6 \\ 24 & 5 & 6 \\ 4 & 1 & -2 \end{vmatrix} = 24$,

$D_2 = \begin{vmatrix} 2 & 18 & 6 \\ 4 & 24 & 6 \\ 3 & 4 & -2 \end{vmatrix} = -12$ and $D_3 = \begin{vmatrix} 2 & 4 & 18 \\ 4 & 5 & 24 \\ 3 & 1 & 4 \end{vmatrix} = 18$. Hence $x_1 = \frac{D_1}{D} = \frac{24}{6} = 4$, $x_2 = \frac{D_2}{D} = -\frac{12}{6} = -2$ and $x_3 = \frac{D_3}{D} = \frac{18}{6} = 3$.

EXAMPLE 2 Show that the system

$$\begin{aligned} x_1 + 3x_2 + 5x_3 + 2x_4 &= 2 \\ -x_2 + 3x_3 + 4x_4 &= 0 \\ 2x_1 + x_2 + 9x_3 + 6x_4 &= -3 \\ 3x_1 + 2x_2 + 4x_3 + 8x_4 &= -1 \end{aligned} \quad (10)$$

has a unique solution and find it by using Cramer's rule.

Solution We saw in Example 3.2.10 on page 95 that

$$|A| = \begin{vmatrix} 1 & 3 & 5 & 2 \\ 0 & -1 & 3 & 4 \\ 2 & 1 & 9 & 6 \\ 3 & 2 & 4 & 8 \end{vmatrix} = 160 \neq 0$$

Thus the system has a unique solution. To find it we compute: $D_1 = -464$; $D_2 = 280$; $D_3 = -56$; $D_4 = 112$. Thus $x_1 = D_1/D = -464/160$, $x_2 = D_2/D = 280/160$, $x_3 = D_3/D = -56/160$, and $x_4 = D_4/D = 112/160$. These solutions can be verified by direct substitution into system (10).

PROBLEMS 3.5 In Problems 1–9 solve the given system by using Cramer's rule.

1. $2x_1 + 3x_2 = -1$
   $-7x_1 + 4x_2 = 47$
2. $3x_1 - x_2 = 0$
   $4x_1 + 2x_2 = 5$
3. $2x_1 + x_2 + x_3 = 6$
   $3x_1 - 2x_2 - 3x_3 = 5$
   $8x_1 + 2x_2 + 5x_3 = 11$
4. $x_1 + x_2 + x_3 = 8$
   $4x_2 - x_3 = -2$
   $3x_1 - x_2 + 2x_3 = 0$
5. $2x_1 + 2x_2 + x_3 = 7$
   $x_1 + 2x_2 - x_3 = 0$
   $-x_1 + x_2 + 3x_3 = 1$
6. $2x_1 + 5x_2 - x_3 = -1$
   $4x_1 + x_2 + 3x_3 = 3$
   $-2x_1 + 2x_2 = 0$
7. $2x_1 + x_2 - x_3 = 4$
   $x_1 + x_3 = 2$
   $-x_2 + 5x_3 = 1$
8. $x_1 + x_2 + x_3 + x_4 = 6$
   $2x_1 - x_3 - x_4 = 4$
   $3x_3 + 6x_4 = 3$
   $x_1 - x_4 = 5$
9. $x_1 - x_4 = 7$
   $2x_2 + x_3 = 2$
   $4x_1 - x_2 = -3$
   $3x_3 - 5x_4 = 2$

*10. Consider the triangle in Figure 3.2.

Figure 3.2
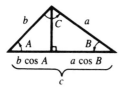

a. Show, using elementary trigonometry, that

$$c \cos A + a \cos C = b$$
$$b \cos A + a \cos B = c$$
$$c \cos B + b \cos C = a$$

b. If the system of part (a) is thought of as a system of three equations in the three unknowns $\cos A$, $\cos B$, and $\cos C$, show that the determinant of the system is nonzero.

c. Use Cramer's rule to solve for $\cos C$.

d. Use part (c) to prove the *law of cosines*: $c^2 = a^2 + b^2 - 2ab \cos C$.

Review Exercises for Chapter 3

In Exercises 1–8 calculate the determinant.

1. $\begin{vmatrix} -1 & 2 \\ 0 & 4 \end{vmatrix}$
2. $\begin{vmatrix} -3 & 5 \\ -7 & 4 \end{vmatrix}$
3. $\begin{vmatrix} 1 & -2 & 3 \\ 0 & 4 & 5 \\ 0 & 0 & 6 \end{vmatrix}$

4. $\begin{vmatrix} 5 & 0 & 0 \\ 6 & 2 & 0 \\ 10 & 100 & 6 \end{vmatrix}$
5. $\begin{vmatrix} 1 & -1 & 2 \\ 3 & 4 & 2 \\ -2 & 3 & 4 \end{vmatrix}$
6. $\begin{vmatrix} 3 & 1 & -2 \\ 4 & 0 & 5 \\ -6 & 1 & 3 \end{vmatrix}$

7. $\begin{vmatrix} 1 & -1 & 2 & 3 \\ 4 & 0 & 2 & 5 \\ -1 & 2 & 3 & 7 \\ 5 & 1 & 0 & 4 \end{vmatrix}$
8. $\begin{vmatrix} 3 & 15 & 17 & 19 \\ 0 & 2 & 21 & 60 \\ 0 & 0 & 1 & 50 \\ 0 & 0 & 0 & -1 \end{vmatrix}$

In Exercises 9–14 use determinants to calculate the inverse, (if one exists).

9. $\begin{pmatrix} -3 & 4 \\ 2 & 1 \end{pmatrix}$
10. $\begin{pmatrix} 3 & -5 & 7 \\ 0 & 2 & 4 \\ 0 & 0 & -3 \end{pmatrix}$
11. $\begin{pmatrix} 1 & -1 & 2 \\ 3 & 1 & 4 \\ 5 & -1 & 8 \end{pmatrix}$

12. $\begin{pmatrix} 1 & 1 & 1 \\ 1 & 0 & 1 \\ 0 & 1 & 1 \end{pmatrix}$
13. $\begin{pmatrix} 2 & 1 & 0 & 0 \\ 0 & -1 & 3 & 0 \\ 1 & 0 & 0 & -2 \\ 3 & 0 & -1 & 0 \end{pmatrix}$
14. $\begin{pmatrix} 3 & -1 & 2 & 4 \\ 1 & 1 & 0 & 3 \\ -2 & 4 & 1 & 5 \\ 6 & -4 & 1 & 2 \end{pmatrix}$

In Exercises 15–20 determine whether the given set of vectors is linearly independent or dependent.

15. $\begin{pmatrix} 1 \\ -1 \end{pmatrix}; \begin{pmatrix} -2 \\ 2 \end{pmatrix}$
16. $(2, 3); (3, 2)$
17. $\begin{pmatrix} 1 \\ 5 \\ 2 \end{pmatrix}; \begin{pmatrix} 3 \\ 0 \\ 4 \end{pmatrix}; \begin{pmatrix} -5 \\ 5 \\ -6 \end{pmatrix}$

18. $(0, 2, 0); (0, 0, 2); (2, 0, 0)$

19. $\begin{pmatrix} 1 \\ 0 \\ 1 \\ 0 \end{pmatrix}; \begin{pmatrix} 1 \\ 1 \\ 0 \\ 0 \end{pmatrix}; \begin{pmatrix} 1 \\ 0 \\ 0 \\ 1 \end{pmatrix}; \begin{pmatrix} 0 \\ 0 \\ 1 \\ 1 \end{pmatrix}$

20. $(3, -1, 1, 4); (7, 1, 0, -2); (0, 2, 2, 3); (-4, 0, 3, 9)$

In Exercises 21–24 solve the system by using Cramer's rule.

21. $\begin{aligned} 2x_1 - x_2 &= 3 \\ 3x_1 + 2x_2 &= 5 \end{aligned}$

22. $\begin{aligned} x_1 - x_2 + x_3 &= 7 \\ 2x_1 - 5x_3 &= 4 \\ 3x_2 - x_3 &= 2 \end{aligned}$

23. $\begin{aligned} 2x_1 + 3x_2 - x_3 &= 5 \\ -x_1 + 2x_2 + 3x_3 &= 0 \\ 4x_1 - x_2 + x_3 &= -1 \end{aligned}$

24. $\begin{aligned} x_1 - x_3 + x_4 &= 7 \\ 2x_2 + 2x_3 - 3x_4 &= -1 \\ 4x_1 - x_2 - x_3 &= 0 \\ -2x_1 + x_2 + 4x_3 &= 2 \end{aligned}$

Answers to Odd-Numbered Problems

CHAPTER 1

Problems 1.2

1. $x_1 = \frac{-13}{5}$, $x_2 = \frac{-11}{5}$; det $= -10$
3. no solutions; det $= 0$
5. $x_1 = \frac{11}{2}$, $x_2 = -30$; det $= -2$
7. infinite number of solutions; $x_2 = \frac{2}{3}x_1$, where x_1 is arbitrary; det $= 0$
9. $x_1 = -1$, $x_2 = 2$; det $= -1$
11. det $= a^2 - b^2$; if $a^2 - b^2 \neq 0$ (i.e., if $a \neq \pm b$), then $x_1 = x_2 = c/(a+b)$. If $a^2 - b^2 = 0$, then $a = \pm b$. If $a \neq 0$ and $a = b$, then there is an infinite number of solutions given by $x_2 = c/a - x_1$. If $a \neq 0$ and $a = -b$, then there are no solutions.
13. det $= -2ab$; so unique solution if both a and b are nonzero
15. $a = b = 0$ and $c \neq 0$ or $d \neq 0$
17. no point of intersection
19. The lines are coincident. Any point of the form $(x, (4x-10)/6)$ is a point of intersection.
21. $(\frac{67}{45}, \frac{2}{15})$ 23. $\sqrt{13}/13$
25. $\sqrt{61}/5$ 27. $\sqrt{5}$
29. Since the slope of the given line L is $-\frac{a}{b}$, the slope of $L_\perp$ is $\frac{b}{a}$. The equation of a line $L_\perp$ perpendicular to L and passing through (x_1, y_1) is given by $\frac{y-y_1}{x-x_1} = \frac{b}{a}$, or $bx - ay = bx_1 - ay_1$. The unique point of intersection of L and $L_\perp$ is found to be
$(x_0, y_0) = \left(\frac{bc - abx_1 + a^2y_1}{a^2 + b^2}, \frac{ac - aby_1 + b^2x_1}{a^2 + b^2}\right)$. Then d is the distance between (x_0, y_0) and (x_1, y_1) and, after some algebra, $d^2 = \frac{1}{(a^2+b^2)^2}$
$\times (a^2c^2 - 2a^2bcy_1 + a^2b^2y_1^2$
$- 2a^3cx_1 + 2a^3bx_1y_1$
$+ a^4x_1^2 + c^2b^2 - 2ab^2cx_1$
$+ a^2b^2x_1^2 - 2b^3cy_1$
$+ 2ab^3x_1y_1 + b^4y_1^2)$.
$= \frac{a^2+b^2}{(a^2+b^2)^2}(c^2 - 2abcy_1$
$+ b^2y_1^2 - 2acx_1 + 2abx_1y_1$
$+ a^2x_1^2) = \frac{1}{(a^2+b^2)}(ax_1 + by_1 - c)^2$. Thus
$d = \frac{|ax_1 + by_1 - c|}{\sqrt{a^2+b^2}}$.
35. Infinite number of solutions; x = no. of cups; y = no. of saucers; solutions are $(x, 240 - \frac{3}{2}x)$.
37. 32 sodas, 128 milk shakes

Problems 1.3

Note: Where there were an infinite number of solutions, we wrote the solutions with the last variable chosen arbitrarily. The solutions can be written in other ways as well.

1. $(2, -3, 1)$
3. $(3 + \frac{2}{9}x_3, \frac{8}{9}x_3, x_3)$, x_3 arbitrary.
5. $(-9, 30, 14)$
7. no solution
9. $(-\frac{4}{5}x_3, \frac{9}{5}x_3, x_3)$, x_3 arbitrary

11. $(-1, \frac{5}{2} + \frac{1}{2}x_3, x_3)$, x_3 arbitrary
13. no solution
15. $(\frac{20}{13} - \frac{4}{13}x_4, \frac{-28}{13} + \frac{3}{13}x_4, \frac{-45}{13} + \frac{9}{13}x_4, x_4)$, arbitrary
17. $(18 - 4x_4, \frac{-15}{2} + 2x_4, -31 + 7x_4, x_4)$, x_4 arbitrary
19. no solution
21. row echelon form
23. reduced row echelon form
25. neither
27. reduced row echelon form
29. neither
31. row echelon form:
$\begin{pmatrix} 1 & -6 \\ 0 & 1 \end{pmatrix}$; reduced row echelon form: $\begin{pmatrix} 1 & 0 \\ 0 & 1 \end{pmatrix}$
33. row echelon form:
$\begin{pmatrix} 1 & -2 & 4 \\ 0 & 1 & -\frac{4}{11} \\ 0 & 0 & 1 \end{pmatrix}$;
reduced row echelon form:
$\begin{pmatrix} 1 & 0 & 0 \\ 0 & 1 & 0 \\ 0 & 0 & 1 \end{pmatrix}$
35. row echelon form:
$\begin{pmatrix} 1 & -\frac{7}{2} \\ 0 & 1 \\ 0 & 0 \end{pmatrix}$;
reduced row echelon form:
$\begin{pmatrix} 1 & 0 \\ 0 & 1 \\ 0 & 0 \end{pmatrix}$
37. $x_1 = 30{,}000 - 5x_3$
$x_2 = x_3 - 5000$
$5000 \leq x_3 \leq 6000$; no

39. No unique solution (2 equations in 3 unknowns); if 200 shares of McDonald's, then 100 shares of Hilton and 300 shares of Eastern.

41. The row echelon form of the augmented matrix representing this system is

$$\begin{pmatrix} 1 & -\frac{1}{2} & \frac{3}{2} & | & a/2 \\ 0 & 1 & \frac{-19}{5} & | & \frac{2}{5}(b - \frac{3}{2}a) \\ 0 & 0 & 0 & | & -2a + 3b + c \end{pmatrix}$$

which is inconsistent if $-2a + 3b + c \neq 0$ or $c \neq 2a - 3b$.

43. $a_{11}a_{22}a_{33} + a_{12}a_{23}a_{31} + a_{13}a_{32}a_{21} - a_{13}a_{22}a_{31} - a_{12}a_{21}a_{33} - a_{11}a_{32}a_{23} \neq 0$

45. (1.900812947, 4.194110816, −11.34851834)

Problems 1.4

1. (0, 0) **3.** (0, 0, 0)
5. $(\frac{1}{6}x_3, \frac{5}{6}x_3, x_3)$, x_3 arbitrary
7. (0, 0)
9. $(-4x_4, 2x_4, 7x_4, x_4)$, x_4 arbitrary.
11. (0, 0) **13.** (0, 0, 0)
15. $k = \frac{95}{11}$

Review Exercises for Chapter 1

1. $(\frac{1}{7}, \frac{10}{7})$ **3.** no solution
5. (0, 0, 0) **7.** $(-\frac{1}{2}, 0, \frac{5}{2})$
9. $(\frac{1}{3}x_3, \frac{7}{3}x_3, x_3)$, x_3 arbitrary
11. no solution
13. (0, 0, 0, 0) **15.** $\sqrt{5}/5$.
17. row echelon form
19. reduced row echelon form
21. row echelon form:

$$\begin{pmatrix} 1 & 4 & -1 \\ 0 & 1 & \frac{5}{4} \end{pmatrix};$$

reduced row echelon form:

$$\begin{pmatrix} 1 & 0 & -6 \\ 0 & 1 & \frac{5}{4} \end{pmatrix}$$

CHAPTER 2
Problems 2.1

1. $\begin{pmatrix} 2 \\ -3 \\ 11 \end{pmatrix}$ **3.** $\begin{pmatrix} -4 \\ 0 \\ 4 \end{pmatrix}$

5. $\begin{pmatrix} -31 \\ 22 \\ -27 \end{pmatrix}$ **7.** $\begin{pmatrix} 0 \\ 0 \\ 0 \end{pmatrix}$

9. $\begin{pmatrix} -11 \\ 11 \\ -10 \end{pmatrix}$ **11.** (1, 2, 5, 7)

13. (−8, 12, 4, 20)
15. (8, −5, 7, −1)
17. (7, 2, 4, 11)
19. (−11, 9, 18, 18)

21. $\mathbf{a} + \mathbf{0} = \begin{pmatrix} a_1 \\ a_2 \\ \vdots \\ a_n \end{pmatrix} + \begin{pmatrix} 0 \\ 0 \\ \vdots \\ 0 \end{pmatrix}$

$= \begin{pmatrix} a_1 + 0 \\ a_2 + 0 \\ \vdots \\ a_n + 0 \end{pmatrix} = \begin{pmatrix} a_1 \\ a_2 \\ \vdots \\ a_n \end{pmatrix} = \mathbf{a}$

23. $\mathbf{a} + \mathbf{b} = \begin{pmatrix} a_1 + b_1 \\ a_2 + b_2 \\ \vdots \\ a_n + b_n \end{pmatrix}$,

$\alpha(\mathbf{a} + \mathbf{b}) = \begin{pmatrix} \alpha(a_1 + b_1) \\ \alpha(a_2 + b_2) \\ \vdots \\ \alpha(a_n + b_n) \end{pmatrix}$

$= \begin{pmatrix} \alpha a_1 + \alpha b_1 \\ \alpha a_2 + \alpha b_2 \\ \vdots \\ \alpha a_n + \alpha b_n \end{pmatrix}$

$= \begin{pmatrix} \alpha a_1 \\ \alpha a_2 \\ \vdots \\ \alpha a_n \end{pmatrix} + \begin{pmatrix} \alpha b_1 \\ \alpha b_2 \\ \vdots \\ \alpha b_n \end{pmatrix}$

$= \alpha \begin{pmatrix} a_1 \\ a_2 \\ \vdots \\ a_n \end{pmatrix} + \alpha \begin{pmatrix} b_1 \\ b_2 \\ \vdots \\ b_n \end{pmatrix}$

$= \alpha \mathbf{a} + \alpha \mathbf{b};$

$(\alpha + \beta)\mathbf{a} = \begin{pmatrix} (\alpha + \beta)a_1 \\ (\alpha + \beta)a_2 \\ \vdots \\ (\alpha + \beta)a_n \end{pmatrix}$

$= \begin{pmatrix} \alpha a_1 + \beta a_1 \\ \alpha a_2 + \beta a_2 \\ \vdots \\ \alpha a_n + \beta a_n \end{pmatrix}$

$= \begin{pmatrix} \alpha a_1 \\ \alpha a_2 \\ \vdots \\ \alpha a_n \end{pmatrix} + \begin{pmatrix} \beta a_1 \\ \beta a_2 \\ \vdots \\ \beta a_n \end{pmatrix}$

$= \alpha \begin{pmatrix} a_1 \\ a_2 \\ \vdots \\ a_n \end{pmatrix} + \beta \begin{pmatrix} a_1 \\ a_2 \\ \vdots \\ a_n \end{pmatrix}$

$= \alpha \mathbf{a} + \beta \mathbf{a};$

$(\alpha\beta)\mathbf{a} = \begin{pmatrix} \alpha\beta a_1 \\ \alpha\beta a_2 \\ \vdots \\ \alpha\beta a_n \end{pmatrix} = \alpha \begin{pmatrix} \beta a_1 \\ \beta a_2 \\ \vdots \\ \beta a_n \end{pmatrix}$

$= \alpha \left[\beta \begin{pmatrix} a_1 \\ a_2 \\ \vdots \\ a_n \end{pmatrix} \right] = \alpha(\beta \mathbf{a})$

25. $\mathbf{d}_1 + \mathbf{d}_2$ represents the combined demand of the two factories for each of the four raw materials needed to produce one unit of each factory's product; $2\mathbf{d}_1$ represents the demand of factory 1 for each of the four raw materials needed to produce two units of its product.

27. $\mathbf{w} = \begin{pmatrix} 3 \\ 0 \\ 5 \end{pmatrix}$

Problems 2.2

1. −14 **3.** 1 **5.** $ac + bd$
7. 51 **9.** $a = 0$ **11.** 4
13. 28 **15.** orthogonal
17. orthogonal **19.** orthogonal

21. all α and β which satisfy $5\alpha + 4\beta = 25$ ($\beta = (25 - 5\alpha)/4$, α arbitrary)

23. (a) (2, 3, 5, 1)

(b) $\begin{pmatrix} 1 \\ \frac{3}{2} \\ \frac{1}{2} \\ 2 \end{pmatrix}$ (c) 11

Problems 2.3

1. $\begin{pmatrix} 3 & 9 \\ 6 & 15 \\ -3 & 6 \end{pmatrix}$ 3. $\begin{pmatrix} 2 & 2 \\ -2 & -1 \\ 6 & -1 \end{pmatrix}$

5. $\begin{pmatrix} 0 & 0 \\ 0 & 0 \\ 0 & 0 \end{pmatrix}$ 7. $\begin{pmatrix} -2 & 4 \\ 7 & 15 \\ -15 & 10 \end{pmatrix}$

9. $\begin{pmatrix} 4 & 10 \\ 17 & 22 \\ -9 & 1 \end{pmatrix}$ 11. $\begin{pmatrix} 0 & 6 \\ 5 & 14 \\ -9 & 9 \end{pmatrix}$

13. $\begin{pmatrix} 1 & -5 & 0 \\ -3 & 4 & -5 \\ -14 & 13 & -1 \end{pmatrix}$

15. $\begin{pmatrix} 1 & 1 & 5 \\ 9 & 5 & 10 \\ 7 & -7 & 3 \end{pmatrix}$

17. $\begin{pmatrix} -1 & -1 & -1 \\ -3 & -3 & -10 \\ -7 & 3 & 5 \end{pmatrix}$

19. $\begin{pmatrix} -1 & -1 & -5 \\ -9 & -5 & -10 \\ -7 & 7 & -3 \end{pmatrix}$

25. $\begin{pmatrix} 1 & 1 & 1 & 0 \\ 1 & 1 & 1 & 0 \\ 1 & 1 & 1 & 1 \\ 0 & 0 & 1 & 1 \end{pmatrix}$

Problems 2.4

1. $\begin{pmatrix} 8 & 20 \\ -4 & 11 \end{pmatrix}$ 3. $\begin{pmatrix} -3 & -3 \\ 1 & 3 \end{pmatrix}$

5. $\begin{pmatrix} 13 & 35 & 18 \\ 20 & 26 & 20 \end{pmatrix}$

7. $\begin{pmatrix} 19 & -17 & 34 \\ 8 & -12 & 20 \\ -8 & -11 & 7 \end{pmatrix}$

9. $\begin{pmatrix} 18 & 15 & 35 \\ 9 & 21 & 13 \\ 10 & 9 & 9 \end{pmatrix}$ 11. $(7 \quad 16)$

13. $\begin{pmatrix} 3 & -2 & 1 \\ 4 & 0 & 6 \\ 5 & 1 & 9 \end{pmatrix}$ 15. $\begin{pmatrix} a & b & c \\ d & e & f \\ g & h & j \end{pmatrix}$

17. If $D = a_{11}a_{22} - a_{12}a_{21}$, then
$$\begin{pmatrix} b_{11} & b_{21} \\ b_{21} & b_{22} \end{pmatrix} = \begin{pmatrix} a_{22}/D & -a_{12}/D \\ -a_{21}/D & a_{11}/D \end{pmatrix}$$

19. (a) 3 in Group 1, 4 in Group 2, 5 in Group 3

(b) $\begin{pmatrix} 2 & 1 & 1 & 0 & 0 \\ 1 & 1 & 0 & 1 & 0 \\ 1 & 0 & 2 & 0 & 1 \end{pmatrix}$

21. (a) $\begin{pmatrix} 80{,}000 & 45{,}000 & 40{,}000 \\ 50 & 20 & 10 \end{pmatrix}$

(b) $\begin{pmatrix} 1 \\ 3 \\ 1 \end{pmatrix}$ (c) Money: 255,000; Shares: 120

23. $\begin{pmatrix} 0 & -8 \\ 32 & 32 \end{pmatrix}$ 25. $\begin{pmatrix} 11 & 38 \\ 57 & 106 \end{pmatrix}$

27. $A^2 = \begin{pmatrix} 0 & 0 & 1 & 0 & 0 \\ 0 & 0 & 0 & 1 & 0 \\ 0 & 0 & 0 & 0 & 1 \\ 0 & 0 & 0 & 0 & 0 \\ 0 & 0 & 0 & 0 & 0 \end{pmatrix}$,

$A^3 = \begin{pmatrix} 0 & 0 & 0 & 1 & 0 \\ 0 & 0 & 0 & 0 & 1 \\ 0 & 0 & 0 & 0 & 0 \\ 0 & 0 & 0 & 0 & 0 \\ 0 & 0 & 0 & 0 & 0 \end{pmatrix}$,

$A^4 = \begin{pmatrix} 0 & 0 & 0 & 0 & 1 \\ 0 & 0 & 0 & 0 & 0 \\ 0 & 0 & 0 & 0 & 0 \\ 0 & 0 & 0 & 0 & 0 \\ 0 & 0 & 0 & 0 & 0 \end{pmatrix}$,

$A^5 = \begin{pmatrix} 0 & 0 & 0 & 0 & 0 \\ 0 & 0 & 0 & 0 & 0 \\ 0 & 0 & 0 & 0 & 0 \\ 0 & 0 & 0 & 0 & 0 \\ 0 & 0 & 0 & 0 & 0 \end{pmatrix}$

29. $PQ = \begin{pmatrix} \frac{11}{90} & \frac{41}{90} & \frac{19}{45} \\ \frac{11}{120} & \frac{71}{120} & \frac{19}{60} \\ \frac{1}{5} & \frac{1}{5} & \frac{3}{5} \end{pmatrix}$;

all entries are nonnegative and
$\frac{11}{90} + \frac{41}{90} + \frac{19}{45} = \frac{11}{120} + \frac{71}{120} + \frac{19}{60}$
$= \frac{1}{5} + \frac{1}{5} + \frac{3}{5} = 1$.

31. Let $P = (p_{ij})$ and $Q = (q_{ij})$ be $k \times k$ probability matrices. Let $PQ = C = (c_{ij})$. The sum of the elements in the mth row of PQ is $c_{m1} + c_{m2} + c_{m3} + \cdots + c_{mk} = p_{m1}q_{11} + p_{m2}q_{21} + p_{m3}q_{31} + \cdots + p_{mk}q_{k1}$
$+ p_{m1}q_{12} + p_{m2}q_{22} + p_{m3}q_{32} + \cdots + p_{mk}q_{k2} + p_{m1}q_{13} + p_{m2}q_{23} + p_{m3}p_{33} + \cdots + p_{mk}q_{k3}$
$\vdots$
$+ p_{m1}q_{1k} + p_{m2}q_{2k} + p_{m3}q_{3k} + \cdots + p_{mk}q_{kk}$

(the elements in parentheses are those of a row of Q, whose sum is 1)

$\overset{\downarrow}{=} p_{m1}(q_{11} + q_{12} + q_{13} + \cdots + q_{1k}) + p_{m2}(q_{21} + q_{22} + q_{23} + \cdots + q_{2k}) + p_{m3}(q_{31} + q_{32} + q_{33} + \cdots + q_{3k}) + \cdots + p_{mk}(q_{k1} + q_{k2} + q_{k3} + \cdots + q_{kk})$
$= p_{m1}(1) + p_{m2}(1) + p_{m3}(1) + \cdots + p_{mk}(1) = 1$.

33. (a) player 2 > player 4 > player 1 > player 3
(b) score = number of games won plus one-half the number of games that were won by each player that this given player beat

35. $A(B + C) = \begin{pmatrix} 1 & 2 & 4 \\ 3 & -1 & 0 \end{pmatrix}$
$\times \begin{pmatrix} 1 & 9 \\ 2 & 11 \\ 10 & 1 \end{pmatrix}$
$= \begin{pmatrix} 45 & 35 \\ 1 & 16 \end{pmatrix}$;

$AB + AC = \begin{pmatrix} 24 & 14 \\ 7 & 17 \end{pmatrix}$
$+ \begin{pmatrix} 21 & 20 \\ -6 & -1 \end{pmatrix}$
$= \begin{pmatrix} 45 & 35 \\ 1 & 16 \end{pmatrix}$

37. 36 **39.** 9840

41. $\frac{13}{3} + \frac{15}{4} + \frac{17}{5} = \frac{689}{60}$

43. $(1^2 + 2^2 + 3^2)(2^3 + 3^3 + 4^3) = 1386$

45. $\sum_{k=0}^{5} (-3)^k$ **47.** $\sum_{k=1}^{n} k^{1/k}$

49. $\sum_{k=0}^{9} \frac{(-1)^{k+1}}{a^k}$

51. $\sum_{k=2}^{7} k^2 \cdot 2k = \sum_{k=2}^{7} 2k^3$

53. $\sum_{i=1}^{3} \sum_{j=1}^{2} a_{ij}$ **55.** $\sum_{k=1}^{5} a_{3k} b_{k2}$

Problems 2.5

1. $\begin{pmatrix} 2 & -1 \\ 4 & 5 \end{pmatrix} \begin{pmatrix} x_1 \\ x_2 \end{pmatrix} = \begin{pmatrix} 3 \\ 7 \end{pmatrix}$

3. $\begin{pmatrix} 3 & 6 & -7 \\ 2 & -1 & 3 \end{pmatrix} \begin{pmatrix} x_1 \\ x_2 \\ x_3 \end{pmatrix} = \begin{pmatrix} 0 \\ 1 \end{pmatrix}$

5. $\begin{pmatrix} 0 & 1 & -1 \\ 1 & 0 & 1 \\ 3 & 2 & 0 \end{pmatrix} \begin{pmatrix} x_1 \\ x_2 \\ x_3 \end{pmatrix} = \begin{pmatrix} 7 \\ 2 \\ -5 \end{pmatrix}$

7. $x_1 + x_2 - x_3 = 7$
$4x_1 - x_2 + 5x_3 = 4$
$6x_1 + x_2 + 3x_3 = 20$

9. $2x_1 \quad\quad + x_3 = 2$
$-3x_1 + 4x_2 \quad\quad = 3$
$\quad\quad 5x_2 + 6x_3 = 5$

11. $x_1 \quad\quad\quad = 2$
$\quad x_2 \quad\quad = 3$
$\quad\quad x_3 \quad = -5$
$\quad\quad\quad x_4 = 6$

13. $6x_1 + 2x_2 + x_3 = 2$
$-2x_1 + 3x_2 + x_3 = 4$
$0x_1 + 0x_2 + 0x_3 = 2$

15. $7x_1 + 2x_2 = 1$
$3x_1 + x_2 = 2$
$6x_1 + 9x_2 = 3$

17. $x_1 = 4 - 2x_2 + 4x_3$; x_2, x_3 arbitrary

19. $x_1 = x_2 = x_3 = 0$

21. $\begin{pmatrix} 1 \\ -\frac{1}{3} \\ \frac{1}{2} \\ 4 \end{pmatrix}$

23. $A = \begin{pmatrix} 2 & 0 & 0 \\ 0 & 4 & 0 \\ 0 & 0 & -5 \end{pmatrix}$, $\mathbf{x} = \begin{pmatrix} x_1 \\ x_2 \\ x_3 \end{pmatrix}$,

$\mathbf{b} = \begin{pmatrix} 3 \\ 5 \\ 2 \end{pmatrix}$; $\mathbf{x} = \begin{pmatrix} \frac{3}{2} \\ \frac{5}{4} \\ -\frac{2}{5} \end{pmatrix}$

Problems 2.6

1. independent

3. dependent;
$-2\begin{pmatrix} 2 \\ -1 \\ 4 \end{pmatrix} + \begin{pmatrix} 4 \\ -2 \\ 8 \end{pmatrix} = \begin{pmatrix} 0 \\ 0 \\ 0 \end{pmatrix}$

5. dependent (from Theorem 2)

7. independent

9. independent

11. independent

13. $ad - bc = 0$ **15.** $\alpha = -\frac{13}{2}$

17. System (7) can be written as

$(*) \quad c_1 \begin{pmatrix} a_{11} \\ a_{21} \\ \vdots \\ a_{m1} \end{pmatrix} + c_2 \begin{pmatrix} a_{12} \\ a_{22} \\ \vdots \\ a_{m2} \end{pmatrix} + \cdots$
$+ c_n \begin{pmatrix} a_{1n} \\ a_{2n} \\ \vdots \\ a_{mn} \end{pmatrix} = \begin{pmatrix} 0 \\ 0 \\ \vdots \\ 0 \end{pmatrix}$

If (7) has even one nontrivial solution, then the columns of A are linearly dependent. If the columns of A are dependent, then there are number $c_1, c_2, \ldots, c_n$ not all zero such that (*) holds.

19. If $0 = c_1 \mathbf{v}_1 + c_2 \mathbf{v}_2 + \cdots + c_k \mathbf{v}_k$, then $0 = c_1 \mathbf{v}_1 + c_2 \mathbf{v}_2 + \cdots + c_k \mathbf{v}_k + 0\mathbf{v}_{k+1} + 0\mathbf{v}_{k+2} + \cdots + 0\mathbf{v}_n$. Since $\mathbf{v}_1, \mathbf{v}_2, \ldots, \mathbf{v}_n$ are independent, we have $c_1 = c_2 = \cdots = c_k = 0$.

21. If $c_1 \mathbf{v}_1 + c_2 \mathbf{v}_2 + c_3 \mathbf{v}_3 = 0$, then $0 = 0 \cdot \mathbf{v}_1 = (c_1 \mathbf{v}_1 + c_2 \mathbf{v}_2 + c_3 \mathbf{v}_3) \cdot \mathbf{v}_1 = c_1(\mathbf{v}_1 \cdot \mathbf{v}_1) + c_2(\mathbf{v}_2 \cdot \mathbf{v}_1) + c_3(\mathbf{v}_3 \cdot \mathbf{v}_1) = c_1 \mathbf{v}_1 \cdot \mathbf{v}_1 + c_2 0 + c_3 0 = c_1 \mathbf{v}_1 \cdot \mathbf{v}_1$. Since $\mathbf{v}_1 \neq 0$, $\mathbf{v}_1 \cdot \mathbf{v}_1 \neq 0$ so we must have $c_1 = 0$. A similar computation shows that $c_2 = c_3 = 0$.

23. $x_2 \begin{pmatrix} -1 \\ 1 \\ 0 \end{pmatrix} + x_3 \begin{pmatrix} -1 \\ 0 \\ 1 \end{pmatrix}$

25. $x_3 \begin{pmatrix} 13 \\ -6 \\ 1 \end{pmatrix}$

Problems 2.7

1. $\begin{pmatrix} 2 & -1 \\ -3 & 2 \end{pmatrix}$ **3.** $\begin{pmatrix} 0 & 1 \\ 1 & 0 \end{pmatrix}$

5. not invertible

7. $\begin{pmatrix} \frac{1}{3} & -\frac{1}{3} & -\frac{1}{3} \\ 0 & \frac{1}{2} & 1 \\ 0 & 0 & -1 \end{pmatrix}$

9. not invertible

11. not invertible

13. $\begin{pmatrix} \frac{7}{3} & -\frac{1}{3} & -\frac{1}{3} & -\frac{2}{3} \\ \frac{4}{9} & -\frac{1}{9} & -\frac{4}{9} & \frac{1}{9} \\ -\frac{1}{9} & -\frac{2}{9} & \frac{1}{9} & \frac{2}{9} \\ -\frac{5}{3} & \frac{2}{3} & \frac{2}{3} & \frac{1}{3} \end{pmatrix}$

15. $\begin{pmatrix} 0 & 1 & 0 & 2 \\ 1 & -1 & -2 & 2 \\ 0 & 1 & 3 & -3 \\ -2 & 2 & 3 & -2 \end{pmatrix}$

17. $(A_1 A_2 \cdots A_m)^{-1} = A_m^{-1} A_{m-1}^{-1} \cdots A_2^{-1} A_1^{-1}$ since $(A_m^{-1} A_{m-1}^{-1} \cdots A_2^{-1} A_1^{-1})(A_1 A_2 \cdots A_{m-1} A_m)$
$= (A_m^{-1} A_{m-1}^{-1} \cdots A_2^{-1})(A_1^{-1} A_1) \times A_2 \cdots A_{m-1} A_m$
$= (A_m^{-1} A_{m-1}^{-1} \cdots A_2^{-1}) \times (A_2 \cdots A_{m-1} A_m) = \cdots = I$

19. $A^{-1} = \dfrac{1}{a_{11} a_{22} - a_{21} a_{12}}$
$\times \begin{pmatrix} a_{22} & -a_{12} \\ -a_{21} & a_{11} \end{pmatrix}$. If $A = \pm I$,

then $A^{-1} = A$. If $a_{11} = -a_{22}$ and $a_{21}a_{12} = 1 - a_{11}^2$, then $a_{11}a_{22} - a_{21}a_{12} = -a_{11}^2 - (1 - a_{11}^2) = -1$. Thus

$$A^{-1} = \begin{pmatrix} -a_{22} & a_{12} \\ a_{21} & -a_{11} \end{pmatrix}$$
$$= \begin{pmatrix} a_{11} & a_{12} \\ a_{21} & a_{22} \end{pmatrix} = A.$$

21. The system $B\mathbf{x} = \mathbf{0}$ has an infinite number of solutions (by Theorem 1.4.1). But if $B\mathbf{x} = \mathbf{0}$, then $AB\mathbf{x} = \mathbf{0}$. Thus, from Theorem 6 (parts [i] and [ii], AB is not invertible.

23. $\begin{pmatrix} \sin\theta & \cos\theta & 0 \\ \cos\theta & -\sin\theta & 0 \\ 0 & 0 & 1 \end{pmatrix}$

is its own inverse (since $\sin^2\theta + \cos^2\theta = 1$).

25. If the ith diagonal component is 0, then in the row reduction of A the ith row is zero so that, by the statement in Step 3(b) on page 64, A is not invertible. Otherwise, if

$$A = \text{diag}(a_1, a_2, \ldots, a_n)$$

then

$$A^{-1} = \text{diag}\left(\frac{1}{a_1}, \frac{1}{a_2}, \ldots, \frac{1}{a_n}\right).$$

27. $\begin{pmatrix} \frac{1}{2} & -\frac{1}{6} & \frac{7}{30} \\ 0 & \frac{1}{3} & -\frac{4}{15} \\ 0 & 0 & \frac{1}{5} \end{pmatrix}$

29. We prove the result for a lower triangular matrix and use Theorem 6, parts (i) and (v). Let

$$A = \begin{pmatrix} a_{11} & 0 & 0 & \cdots & 0 \\ a_{21} & a_{22} & 0 & \cdots & 0 \\ a_{31} & a_{32} & a_{33} & \cdots & 0 \\ \vdots & \vdots & \vdots & & \vdots \\ a_{n1} & a_{n2} & a_{n3} & \cdots & a_{nn} \end{pmatrix}$$

Taking a linear combination of the columns, we suppose that

$$\begin{pmatrix} 0 \\ 0 \\ \vdots \\ 0 \end{pmatrix} = c_1 \begin{pmatrix} a_{11} \\ a_{21} \\ \vdots \\ a_{n1} \end{pmatrix} + c_2 \begin{pmatrix} 0 \\ a_{22} \\ \vdots \\ a_{n2} \end{pmatrix}$$
$$+ c_3 \begin{pmatrix} 0 \\ 0 \\ a_{33} \\ \vdots \\ a_{n3} \end{pmatrix} + c_n \begin{pmatrix} 0 \\ 0 \\ 0 \\ \vdots \\ a_{nn} \end{pmatrix},$$

or $a_{11}c_1 = 0$
$a_{21}c_1 + a_{22}c_2 = 0$
$a_{31}c_1 + a_{32}c_2 + a_{33}c_3 = 0$
$\vdots \quad \vdots \quad \vdots \quad \vdots$
$a_{n1}c_1 + a_{n2}c_2 + a_{n3}c_3 + \cdots + a_{nn}c_n = 0.$

Suppose that none of the diagonal elements is zero. Then, from the first equation, $c_1 = 0$. Similarly, from the second equation, $c_2 = 0$, etc. Thus the rows of A are linearly independent so that A is invertible. Suppose that $a_{ii} = 0$. Then the first i equations read

$a_{11}c_1 = 0$
$a_{21}c_1 + a_{22}c_2 = 0$
$\vdots$
$a_{i-1,1}c_1 + a_{i-1,2}c_2 + \cdots + a_{i-1,i-1}c_{i-1} = 0$
$a_{i1}c_1 + a_{i2}c_2 + \cdots + a_{i,i-1}c_{i-1} = 0$

This is a system of i equations in the $i - 1$ unknowns $c_1, c_2, \ldots, c_{i-1}$. By Theorem 1.4.1, the system has an infinite number of solutions. Thus the first i rows of A are linearly dependent so that A is not invertible.

31. any nonzero multiple of $(1, 2)$.

33. 3 chairs and 2 tables

35. 4 units of A and 5 units of B

37. (a) $A = \begin{pmatrix} 0.293 & 0 & 0 \\ 0.014 & 0.207 & 0.017 \\ 0.044 & 0.010 & 0.216 \end{pmatrix}$;

$I - A$
$= \begin{pmatrix} 0.707 & 0 & 0 \\ -0.014 & 0.793 & -0.017 \\ -0.044 & -0.010 & 0.784 \end{pmatrix}$

(b) $\begin{pmatrix} 195492.2207 \\ 25932.85859 \\ 13580.33966 \end{pmatrix}$

39. $\begin{pmatrix} 1 & \frac{1}{2} \\ 0 & 1 \end{pmatrix}$; yes

41. $\begin{pmatrix} 1 & \frac{2}{3} & \frac{1}{3} \\ 0 & 1 & 1 \\ 0 & 0 & 1 \end{pmatrix}$; yes

43. $\begin{pmatrix} 1 & -\frac{1}{2} & 2 \\ 0 & 1 & -14 \\ 0 & 0 & 0 \end{pmatrix}$; no

45. $\begin{pmatrix} 1 & 0 & 2 & 3 \\ 0 & 1 & 2 & 7 \\ 0 & 0 & 1 & \frac{10}{7} \\ 0 & 0 & 0 & 0 \end{pmatrix}$; no

Problems 2.8

1. $\begin{pmatrix} -1 & 6 \\ 4 & 5 \end{pmatrix}$ **3.** $\begin{pmatrix} 2 & -1 & 1 \\ 3 & 2 & 4 \end{pmatrix}$

5. $\begin{pmatrix} 1 & -1 & 1 \\ 2 & 0 & 5 \\ 3 & 4 & 5 \end{pmatrix}$ **7.** $\begin{pmatrix} 1 & 0 \\ 0 & 1 \\ 1 & 0 \\ 0 & 1 \end{pmatrix}$

9. $\begin{pmatrix} a & d & g \\ b & e & h \\ c & f & j \end{pmatrix}$

11. $[(A + B)^t]_{ij} = (A + B)_{ji} = a_{ji} + b_{ji} = (A^t)_{ij} + (B^t)_{ij}$. Thus the ijth component of $(A + B)^t$ equals the ijth component of A^t plus the ijth component of B^t.

15. If A is $m \times n$, then A^t is $n \times m$ and AA^t is $m \times m$. Also, $(AA^t)^t = (A^t)^t A^t = AA^t$.

17. If A is upper triangular and $B = A^t$, then $b_{ij} = a_{ji} = 0$ if $j > i$. Thus B is lower triangular.

19. $(A+B)^t = A^t + B^t = -A - B = -(A+B)$

21. $(AB)^t = B^t A^t = (-B)(-A) = (-1)^2 BA = BA$

Chapter 2—Review

1. $\begin{pmatrix} 4 \\ 6 \\ 5 \end{pmatrix}$ **3.** $\begin{pmatrix} 6 \\ 0 \\ -7 \end{pmatrix}$ **5.** -3

7. 207 **9.** $\begin{pmatrix} -6 & 3 \\ 0 & 12 \\ 6 & 9 \end{pmatrix}$

11. $\begin{pmatrix} 16 & 2 & 3 \\ -20 & 10 & -1 \\ -36 & 8 & 16 \end{pmatrix}$

13. $\begin{pmatrix} 17 & 39 & 41 \\ 14 & 20 & 42 \end{pmatrix}$

15. $\begin{pmatrix} 9 & 10 \\ 30 & 32 \end{pmatrix}$

17. $A(BC) = \begin{pmatrix} 25 & 74 \\ 132 & 222 \end{pmatrix} = (AB)C$

19. dependent; $\begin{pmatrix} 4 \\ 6 \end{pmatrix} = 2 \begin{pmatrix} 2 \\ 3 \end{pmatrix}$

21. dependent;
$2\begin{pmatrix} 1 \\ -4 \\ 2 \end{pmatrix} - \begin{pmatrix} 0 \\ 2 \\ -1 \end{pmatrix} - \begin{pmatrix} 2 \\ -10 \\ 5 \end{pmatrix} = \begin{pmatrix} 0 \\ 0 \\ 0 \end{pmatrix}$

23. $\begin{pmatrix} 1 & \frac{3}{2} \\ 0 & 1 \end{pmatrix}$; inverse is $\begin{pmatrix} \frac{4}{11} & -\frac{3}{11} \\ \frac{1}{11} & \frac{2}{11} \end{pmatrix}$

25. $\begin{pmatrix} 1 & 2 & 0 \\ 0 & 1 & \frac{1}{3} \\ 0 & 0 & 1 \end{pmatrix}$; inverse is $\begin{pmatrix} -\frac{1}{4} & \frac{1}{4} & \frac{1}{4} \\ \frac{5}{8} & -\frac{1}{8} & -\frac{1}{8} \\ \frac{1}{8} & -\frac{5}{8} & \frac{3}{8} \end{pmatrix}$

27. $\begin{pmatrix} 1 & 0 & 2 \\ 0 & 1 & 1 \\ 0 & 0 & 1 \end{pmatrix}$; inverse is $\begin{pmatrix} \frac{5}{6} & \frac{2}{3} & -2 \\ \frac{1}{3} & \frac{2}{3} & -1 \\ -\frac{1}{6} & -\frac{1}{3} & 1 \end{pmatrix}$

29. $\begin{pmatrix} 1 & 2 & 0 \\ 2 & 1 & -1 \\ 3 & 1 & 1 \end{pmatrix} \begin{pmatrix} x_1 \\ x_2 \\ x_3 \end{pmatrix} = \begin{pmatrix} 3 \\ -1 \\ 7 \end{pmatrix}$;

A^{-1} is given in Problem 25; $x_1 = \frac{3}{4}$, $x_2 = \frac{9}{8}$, $x_3 = \frac{29}{8}$

31. $\begin{pmatrix} 2 & -1 \\ 3 & 0 \\ 1 & 2 \end{pmatrix}$; neither

33. $\begin{pmatrix} 2 & 3 & 1 \\ 3 & -6 & -5 \\ 1 & -5 & 9 \end{pmatrix}$; symmetric

35. $\begin{pmatrix} 1 & -1 & 4 & 6 \\ -1 & 2 & 5 & 7 \\ 4 & 5 & 3 & -8 \\ 6 & 7 & -8 & 9 \end{pmatrix}$; symmetric

CHAPTER 3

Problems 3.1

1. -10 **3.** 47 **5.** 4 **7.** 56
9. 274

11. Let $A = \begin{pmatrix} a_{11} & 0 & 0 & \cdots & 0 \\ 0 & a_{22} & 0 & \cdots & 0 \\ 0 & 0 & a_{33} & \cdots & 0 \\ \vdots & \vdots & \vdots & & \vdots \\ 0 & 0 & 0 & \cdots & a_{nn} \end{pmatrix}$

and

$B = \begin{pmatrix} b_{11} & 0 & 0 & \cdots & 0 \\ 0 & b_{22} & 0 & \cdots & 0 \\ 0 & 0 & b_{33} & \cdots & 0 \\ \vdots & \vdots & \vdots & & \vdots \\ 0 & 0 & 0 & \cdots & b_{nn} \end{pmatrix}$.

Then $\det A = a_{11} a_{22} a_{33} \cdots a_{nn}$, $\det B = b_{11} b_{22} b_{33} \cdots b_{nn}$,

$AB = \begin{pmatrix} a_{11}b_{11} & 0 & 0 & \cdots & 0 \\ 0 & a_{22}b_{22} & 0 & \cdots & 0 \\ 0 & 0 & a_{33}b_{33} & \cdots & 0 \\ \vdots & \vdots & \vdots & & \vdots \\ 0 & 0 & 0 & \cdots & a_{nn}b_{nn} \end{pmatrix}$

and
$\det AB = (a_{11}b_{11})(a_{22}b_{22}) \times (a_{33}b_{33}) \cdots (a_{nn}b_{nn})$
$= (a_{11}a_{22}a_{33} \cdots a_{nn}) \times (b_{11}b_{22}b_{33} \cdots b_{nn})$
$= \det A \det B$.

13. Almost any example will work. For instance,

$\det \begin{pmatrix} 1 & 0 \\ 0 & 1 \end{pmatrix} = 1$, but

$\det \begin{pmatrix} 1 & 0 \\ 0 & 0 \end{pmatrix} + \det \begin{pmatrix} 0 & 0 \\ 0 & 1 \end{pmatrix}$
$= 0 + 0 \neq 1$.

As another example, let
$A = \begin{pmatrix} 1 & 2 \\ 3 & 4 \end{pmatrix}$ and $B = \begin{pmatrix} 5 & 6 \\ 7 & 8 \end{pmatrix}$; then $(A+B) = \begin{pmatrix} 6 & 8 \\ 10 & 12 \end{pmatrix}$, $\det A = -2$, $\det B = -2$, and $\det(A+B) = -8 \neq \det A + \det B$.

15. Let $A = \begin{pmatrix} a_{11} & 0 & \cdots & 0 \\ a_{21} & a_{22} & \cdots & 0 \\ \vdots & \vdots & & \vdots \\ a_{n1} & a_{n2} & \cdots & a_{nn} \end{pmatrix}$.

Then, continually expanding in the first row, we obtain

$\det A = a_{11} \begin{vmatrix} a_{22} & 0 & \cdots & 0 \\ a_{32} & a_{33} & \cdots & 0 \\ \vdots & \vdots & & \vdots \\ a_{n2} & a_{n3} & \cdots & a_{nn} \end{vmatrix}$

$= a_{11} a_{22} \begin{vmatrix} a_{33} & 0 & \cdots & 0 \\ a_{43} & a_{44} & \cdots & 0 \\ \vdots & \vdots & & \vdots \\ a_{n3} & a_{n4} & \cdots & a_{nn} \end{vmatrix}$

$= \cdots = a_{11} a_{22} a_{33} \cdots a_{n-2} \begin{vmatrix} a_{n-1,n-1} & 0 \\ a_{n,n-1} & a_{nn} \end{vmatrix}$

$= a_{11} a_{22} a_{33} \cdots a_{n-2,n-2} a_{n-1,n-1} a_{nn}$.

17. Let $\mathbf{u}_1 = \begin{pmatrix} u_{11} \\ u_{12} \end{pmatrix}$, $\mathbf{u}_2 = \begin{pmatrix} u_{21} \\ u_{22} \end{pmatrix}$

and $A = \begin{pmatrix} a_{11} & a_{12} \\ a_{21} & a_{22} \end{pmatrix}$. Then

$\mathbf{v}_1 = \begin{pmatrix} a_{11} & a_{12} \\ a_{21} & a_{22} \end{pmatrix} \begin{pmatrix} u_{11} \\ u_{12} \end{pmatrix}$

$= \begin{pmatrix} a_{11}u_{11} + a_{12}u_{12} \\ a_{21}u_{11} + a_{22}u_{12} \end{pmatrix}$,

$\mathbf{v}_2 = \begin{pmatrix} a_{11} & a_{12} \\ a_{21} & a_{22} \end{pmatrix} \begin{pmatrix} u_{21} \\ u_{22} \end{pmatrix}$

$= \begin{pmatrix} a_{11}u_{21} + a_{12}u_{22} \\ a_{21}u_{21} + a_{22}u_{22} \end{pmatrix}$.

Let $A_\mathbf{v}$ denote the area generated by $\mathbf{v}_1$ and $\mathbf{v}_2$. Then, applying the result of Problem 16,

$A_\mathbf{v} = |(a_{11}u_{11} + a_{12}u_{12})$
$\quad \times (a_{21}u_{21} + a_{22}u_{22})$
$\quad - (a_{11}u_{21} + a_{12}u_{22})$
$\quad \times (a_{21}u_{11} + a_{22}u_{12})|$

$= |a_{21}a_{11}u_{21}u_{11}$
$\quad + a_{21}a_{12}u_{21}u_{12}$
$\quad + a_{21}a_{11}u_{11}u_{22}$
$\quad + a_{22}a_{12}u_{12}u_{22}$
$\quad - a_{21}a_{11}u_{21}u_{11}$
$\quad - a_{22}a_{11}u_{21}u_{12}$
$\quad - a_{21}a_{12}u_{11}u_{22}$
$\quad - a_{22}a_{12}u_{12}u_{22}|$

$= |u_{11}u_{22} - u_{12}u_{21}|$
$\quad \times |a_{22}a_{11} - a_{21}a_{12}|$

$= A_\mathbf{u} |\det A|$

where $A_\mathbf{u}$ is the area generated by $\mathbf{u}_1$ and $\mathbf{u}_2$, again using the result of Problem 16.

Problems 3.2

1. 28 **3.** 2 **5.** 32 **7.** -36
9. -260 **11.** -183 **13.** 24
15. -296 **17.** -138
19. $abcde$ **21.** -8 **23.** 16
25. -16 **27.** -16
29. Proof by induction: true for $n = 2$ since

$\begin{vmatrix} 1+x_1 & x_2 \\ x_1 & 1+x_2 \end{vmatrix}$

$= (1+x_1)(1+x_2) - x_1x_2$
$= 1 + x_1 + x_2$.

Assume true for $n = k$. That is,

$\begin{vmatrix} 1+x_1 & x_2 & x_3 & \cdots & x_k \\ x_1 & 1+x_2 & x_3 & \cdots & x_k \\ x_1 & x_2 & 1+x_3 & \cdots & x_k \\ \vdots & \vdots & \vdots & & \vdots \\ x_1 & x_2 & x_3 & \cdots & 1+x_k \end{vmatrix}$

$= 1 + x_1 + x_2 + \cdots + x_k$.

Then, for $n = k+1$,

$\begin{vmatrix} 1+x_1 & x_2 & x_3 & \cdots & x_k & x_{k+1} \\ x_1 & 1+x_2 & x_3 & \cdots & x_k & x_{k+1} \\ x_1 & x_2 & 1+x_3 & \cdots & x_k & x_{k+1} \\ \vdots & \vdots & \vdots & & \vdots & \vdots \\ x_1 & x_2 & x_3 & \cdots & x_k & 1+x_{k+1} \end{vmatrix}$

(using Property 3 in the first column)

$= \begin{vmatrix} 1 & x_2 & x_3 & \cdots & x_k & x_{k+1} \\ 0 & 1+x_2 & x_3 & \cdots & x_k & x_{k+1} \\ 0 & x_2 & 1+x_3 & \cdots & x_k & x_{k+1} \\ \vdots & \vdots & \vdots & & \vdots & \vdots \\ 0 & x_2 & x_3 & \cdots & x_k & 1+x_{k+1} \end{vmatrix}$ ①

$+ \begin{vmatrix} x_1 & x_2 & x_3 & \cdots & x_k & x_{k+1} \\ x_1 & 1+x_2 & x_3 & \cdots & x_k & x_{k+1} \\ x_1 & x_2 & 1+x_3 & \cdots & x_k & x_{k+1} \\ \vdots & \vdots & \vdots & & \vdots & \vdots \\ x_1 & x_2 & x_3 & \cdots & x_k & 1+x_{k+1} \end{vmatrix}$ ②

But, expanding det ① in its first column, we have

$\det ① = \begin{vmatrix} 1+x_2 & x_3 & \cdots & x_k & x_{k+1} \\ x_2 & 1+x_3 & \cdots & x_k & x_{k+1} \\ \vdots & \vdots & & \vdots & \vdots \\ x_2 & x_3 & \cdots & x_k & 1+x_{k+1} \end{vmatrix}$

$= 1 + x_2 + x_3 + \cdots + x_{k+1}$

by the induction assumption (since ① is a $k \times k$ determinant). To evaluate det ②, subtract the first row from all other rows:

$\det ② = \begin{vmatrix} x_1 & x_2 & x_3 & \cdots & x_k & x_{k+1} \\ 0 & 1 & 0 & \cdots & 0 & 0 \\ 0 & 0 & 1 & \cdots & 0 & 0 \\ \vdots & \vdots & \vdots & & \vdots & \vdots \\ 0 & 0 & 0 & \cdots & 0 & 1 \end{vmatrix} = x_1.$

Adding det ① and det ② completes the proof.

31. If n is odd, $\det A = -\det A$ so that $2 \det A = 0$ and $\det A = 0$.

33. $\frac{1}{2} \begin{vmatrix} 1 & x_1 & y_1 \\ 1 & x_2 & y_2 \\ 1 & x_3 & y_3 \end{vmatrix}$

$= \frac{1}{2} \begin{vmatrix} x_2 - x_1 & x_3 - x_1 \\ y_2 - y_1 & y_3 - y_1 \end{vmatrix}$.

Look at the figures below.

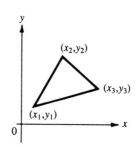

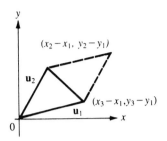

The area A of the triangle is half the area of the parallelogram generated by the vectors $\mathbf{u}_1$ and $\mathbf{u}_2$ which, by the result of Problem 3.1.16, is given by

$$A = \pm \frac{1}{2} \begin{vmatrix} x_2 - x_1 & x_3 - x_1 \\ y_2 - y_1 & y_3 - y_1 \end{vmatrix}.$$

35. $D_3 = \begin{vmatrix} 1 & 1 & 1 \\ a_1 & a_2 & a_3 \\ a_1^2 & a_2^2 & a_3^2 \end{vmatrix}$

$= \begin{vmatrix} 1 & 0 & 0 \\ a_1 & a_2 - a_1 & a_3 - a_1 \\ a_1^2 & a_2^2 - a_1^2 & a_3^2 - a_1^2 \end{vmatrix}$

$= \begin{vmatrix} a_2 - a_1 & a_3 - a_1 \\ (a_2 + a_1)(a_2 - a_1) & (a_3 + a_1)(a_3 - a_1) \end{vmatrix}$

$= (a_2 - a_1)(a_3 - a_1) \begin{vmatrix} 1 & 1 \\ a_2 + a_1 & a_3 + a_1 \end{vmatrix}$

$= (a_2 - a_1)(a_3 - a_1) \times (a_3 - a_2)$

37. (a) $D_n = \begin{vmatrix} 1 & 1 & \cdots & 1 \\ a_1 & a_2 & \cdots & a_n \\ a_1^2 & a_2^2 & \cdots & a_n^2 \\ \vdots & \vdots & & \vdots \\ a_1^{n-1} & a_1^{n-1} & \cdots & a_n^{n-1} \end{vmatrix}$

(b) We prove this by induction. The result is true for $n = 3$ by the result of Problem 35. We assume it true for $n = k$. Now

$D_{k+1} = \begin{vmatrix} 1 & 1 & \cdots & 1 & 1 \\ a_1 & a_2 & \cdots & a_k & a_{k+1} \\ a_1^2 & a_2^2 & \cdots & a_k^2 & a_{k+1}^2 \\ \vdots & \vdots & & \vdots & \vdots \\ a_1^{k-1} & a_2^{k-1} & \cdots & a_k^{k-1} & a_{k+1}^{k-1} \\ a_1^k & a_2^k & \cdots & a_k^k & a_{k+1}^k \end{vmatrix}.$

We subtract the first column from each of the other k columns:

$D_{k+1} = \begin{vmatrix} 1 & 0 & \cdots \\ a_1 & a_2 - a_1 & \cdots \\ a_1^2 & a_2^2 - a_1^2 & \cdots \\ \vdots & \vdots & \\ a_1^{k-1} & a_2^{k-1} - a_1^{k-1} & \cdots \\ a_1^k & a_2^k - a_1^k & \cdots \end{vmatrix}$

$= \begin{vmatrix} a_2 - a_1 & a_3 - a_1 & \cdots & a_{k+1} - a_1 \\ a_2^2 - a_1^2 & a_3^2 - a_1^2 & \cdots & a_{k+1}^2 - a_1^2 \\ \vdots & \vdots & & \vdots \\ a_2^{k-1} - a_1^{k-1} & a_3^{k-1} - a_1^{k-1} & \cdots & a_{k+1}^{k-1} - a_1^{k-1} \\ a_2^k - a_1^k & a_3^k - a_1^k & \cdots & a_{k+1}^k - a_1^k \end{vmatrix}$

Now $a_2^k - a_1^k = (a_2 - a_1) \times (a_2^{k-1} + a_2^{k-2}a_1 + a_2^{k-3}a_1^2 + \cdots + a_2^2 a_1^{k-3} + a_2 a_1^{k-2} + a_1^{k-1})$, and $a_2^{k-1} - a_1^{k-1} = (a_2 - a_1) \times (a_2^{k-2} + a_2^{k-3}a_1 + \cdots + a_2 a_1^{k-4} + a_2 a_1^{k-3} + a_1^{k-2})$. Note that if the terms in the second factor of the last expression are multiplied by a_1, and then subtracted from the second factor of $a_2^k - a_1^k$, only the term a_2^{k-1} remains. Thus we **(i)** expand the last determinant obtained above in the first row, **(ii)** factor $a_{j-1} - a_1$ from the jth column for $1 \leq j \leq k$, and **(iii)**

multiply the ℓth row by a_1 and subtract it from the $(\ell + 1)$st row, for $\ell = k - 1, k - 2, \ldots, 3, 2$ in succession. This yields

$D_{k+1} = (a_2 - a_1)$
$\times (a_3 - a_1) \cdots (a_{k+1} - a_1)$

$\times \begin{vmatrix} 1 & 1 & \cdots & 1 & 1 \\ a_2 & a_3 & \cdots & a_k & a_{k+1} \\ a_2^2 & a_3^2 & \cdots & a_k^2 & a_{k+1}^2 \\ \vdots & \vdots & & \vdots & \vdots \\ a_2^{k-1} & a_3^{k-1} & \cdots & a_k^{k-1} & a_{k+1}^{k-1} \\ a_2^k & a_3^k & \cdots & a_k^k & a_{k+1}^k \end{vmatrix}$

$= \prod_{j=2}^{k+1} (a_j - a_1) \prod_{\substack{i=2 \\ j>i}}^{k+1} (a_j - a_i)$

(from the induction assumption since the last determinant is

$k \times k) = \prod_{\substack{i=1 \\ j>i}}^{k+1} (a_j - a_i);$

this completes the proof.

Problems 3.3

1. $a_{1k} A_{1k}$ is the only term in the expansion in the first column of A involving the component a_{1k}. But

$a_{1k} A_{1k} = a_{1k}(-1)^{1+k} |M_{1k}|.$

If we expand $|M_{1k}|$ about its lth column for $l \neq k$, a term in the expansion takes the form a_{il} (cofactor of a_{il} in M_{1k}). But this is the only occurrence of a_{il} in the expansion of M_{1k} since the other terms have the form a_{jl} (cofactor of a_{jl} in M_{1k}), which deletes the column corresponding to the lth column of A, and a_{il} is in the lth column. Therefore $a_{1k} A_{1k} = (-1)^{1+h} a_{1k} a_{il} \cdot$ (cofactor of a_{il} in m_{1k}).

3. Expand $|A|$ about its kth column. A term is $a_{ik} A_{ik}$ and this is the only occurrence of a_{ij} in the expansion of $|A|$. Now $A_{ik} = (-1)^{i+k} |M_{ik}|$, and if this is expanded in the lth column (for $l \neq k$), the only term in the expansion containing a_{jl} is $a_{jl} \cdot$ (cofactor of a_{jl} in M_{ik}) for the same reason as in Problem 1. Thus the only occurrence of $a_{ij}a_{jl}$ is $(-1)^{i+k}a_{ik}a_{jl} \cdot$ (cofactor of a_{jl} in M_{ik}).

5. -6

Problems 3.4

1. $\begin{pmatrix} \frac{1}{2} & -\frac{1}{2} \\ -\frac{1}{4} & \frac{3}{4} \end{pmatrix}$ **3.** $\begin{pmatrix} 0 & 1 \\ 1 & 0 \end{pmatrix}$

5. $\begin{pmatrix} \frac{1}{3} & -\frac{1}{4} & -\frac{1}{6} \\ 0 & \frac{1}{4} & \frac{1}{2} \\ 0 & \frac{1}{4} & -\frac{1}{2} \end{pmatrix}$

7. $\begin{pmatrix} 0 & 1 & -1 \\ 2 & -2 & -1 \\ -1 & 1 & 1 \end{pmatrix}$

9. not invertible

11. $\begin{pmatrix} \frac{7}{3} & -\frac{1}{3} & -\frac{1}{3} & -\frac{2}{3} \\ \frac{4}{9} & -\frac{1}{9} & -\frac{4}{9} & \frac{1}{9} \\ -\frac{1}{9} & -\frac{2}{9} & \frac{1}{9} & \frac{2}{9} \\ -\frac{5}{3} & \frac{2}{3} & \frac{2}{3} & \frac{1}{3} \end{pmatrix}$

13. independent

15. independent

17. Follows from the fact that $\det A^t = \det A$.

19. $A^{-1} = \begin{pmatrix} \frac{1}{14} & \frac{1}{14} & \frac{9}{28} \\ -\frac{5}{7} & \frac{2}{7} & -\frac{3}{14} \\ \frac{1}{14} & \frac{1}{14} & -\frac{5}{28} \end{pmatrix}$,

$\det A = -28$, $\det A^{-1} = -\frac{1}{28}$

21. no inverse if α is any real number

Problems 3.5

1. $x_1 = -5$, $x_2 = 3$

3. $x_1 = 2$, $x_2 = 5$, $x_3 = -3$

5. $x_1 = \frac{45}{13}$, $x_2 = -\frac{11}{13}$, $x_3 = \frac{23}{13}$

7. $x_1 = \frac{3}{2}$, $x_2 = \frac{3}{2}$, $x_3 = \frac{1}{2}$

9. $x_1 = \frac{21}{29}$, $x_2 = \frac{171}{29}$, $x_3 = -\frac{284}{29}$, $x_4 = -\frac{182}{29}$

Chapter 3—Review

1. -4 **3.** 24 **5.** 60 **7.** 34

9. $\begin{pmatrix} -\frac{1}{11} & \frac{4}{11} \\ \frac{2}{11} & \frac{3}{11} \end{pmatrix}$

11. not invertible

13. $\begin{pmatrix} \frac{1}{11} & \frac{1}{11} & 0 & \frac{3}{11} \\ \frac{9}{11} & -\frac{2}{11} & 0 & -\frac{6}{11} \\ \frac{3}{11} & \frac{3}{11} & 0 & -\frac{2}{11} \\ \frac{1}{22} & \frac{1}{22} & -\frac{1}{2} & \frac{1}{22} \end{pmatrix}$

15. dependent **17.** dependent

19. independent

21. $x_1 = \frac{11}{7}$, $x_2 = \frac{1}{7}$

23. $x_1 = \frac{1}{4}$, $x_2 = \frac{5}{4}$, $x_3 = -\frac{3}{4}$

Index

addition
 of matrices, 38
 of vectors, 29
adjoint of a matrix, 106
associative law
 of matrix addition, 40
 of matrix multiplication, 44
 of vector addition, 31
augmented matrix, 8
back substitution, 14
coefficient matrix, 8, 14
cofactor, 85
 expansion by—s, 85
column of a matrix, 37
commutative law
 of matrix addition, 40
 of scalar product, 35
 of vector addition, 31
complex number, 29
component
 of a matrix, 37
 of a vector, 28
consistent system of equations, 16
Cramer's rule, 113
determinant
 of a 2 × 2 matrix, 65
 of a 2 × 2 system, 4
 of a 3 × 3 matrix, 82
 of an n × n matrix, 85
 Vandermonde, 109
diagonal matrix, 75, 86
diagonal of a matrix, 59
direct contact matrix, 43
distributive law
 for matrix multiplication, 46
 for scalar multiplication, 40
 for scalar product, 35
 for vector addition, 31
dot product, 34

elementary row operations, 10
equivalent system of equations, 8
error round off, 16
expansion by cofactors, 85
Gaussian elimination, 15
Gauss-Jordan elimination, 8, 15
graph, 41
homogeneous system of equations, 22, 53
identity matrix, 59
inconsistent system of equations, 16
indirect contact matrix, 44
inverse of a matrix, 60
invertible matrix, 60
law of cosines, 116
Leontief
 input-output model, 17
 matrix, 69
line
 slope of, 1
 straight, 1
linear
 dependence, 53
 independence, 53
lines
 parallel, 1
 perpendicular, 1
lower triangular
 matrix, 75, 86
mathematical induction, 98
matrices
 addition of, 38
 equal, 37
 product of, 41
 scalar multiplication of, 39
matrix, 8, 37
 adjoint of, 106
 augmented, 8
 coefficient, 8

cofactor of, 85
column of, 37
component of, 37
diagonal, 75, 86
direct contact, 43
identity, 59
inverse of, 60
Leontief, 69
lower triangular, 75, 86
orthogonal, 102
probability, 48
row of, 37
skew-symmetric, 80, 102
square, 37
symmetric, 79
technology, 69
transpose of, 78
triangular, 75, 86
upper triangular, 75, 86
zero, 37
minor, 85
 second-order, 104
nontrivial solution, 22
ordered set, 27
orthogonal
 matrix, 102
 vectors, 36
probability matrix, 48
product of matrices, 42
quaternions, 27
reduced row echelon form, 13
row echelon form, 13
 reduced, 13
row equivalent matrices, 68
row of a matrix, 37
row reduction, 10
 notation for, 10
scalar multiplication
 of matrices, 39

of vectors, 30
scalar product, 33
skew-symmetric matrix, 80, 102
slope of a line, 1
solution to a system of equations, 2
square matrix, 37
Summing Up Theorem, 4, 72, 110
system of linear equations, 2, 6
 consistent, 16
 homogeneous, 22
 inconsistent, 16
 and matrices, 50
technology matrix, 69
transpose of a matrix, 78
triangular matrix, 75, 86
trivial solution, 22
upper triangular matrix, 75, 86
Vandermonde determinant, 102
vector, 27
 column, 28
 component of, 28
 demand, 34
 n-vector, 27, 28
 price, 34
 row, 27
 zero, 28
vectors
 addition of, 29
 dot product of, 34
 equal, 29
 orthogonal, 36
 scalar product of, 33
zero
 equation, 16
 matrix, 37
 solution, 22
 vector, 28